Calcium and Your Health

Calcium and Your Health

by Takuo Fujita, M.D.

Japan Publications, Inc.

Published by JAPAN PUBLICATIONS, INC., Tokyo and New York

Distributors:
UNITED STATES: *Kodansha International/USA, Ltd., through Harper & Row, Publishers, Inc., 10 East 53rd Street, New York, New York 10022.* SOUTH AMERICA: *Harper & Row, Publishers, Inc., International Department.* CANADA: *Fitzhenry & Whiteside Ltd., 195 Allstate Parkway, Markham, Ontario, L3R 4T8.* MEXICO AND CENTRAL AMERICA: *HARLA S. A. de C. V., Apartado 30–546, Mexico 4, D. F.* BRITISH ISLES: *International Book Distributors Ltd., 66 Wood Lane End, Hemel Hempstead, Herts HP2 4RG.* EUROPEAN CONTINENT (except Germany): *PBD Proost & Brandt Distribution bv, Strijkviertel 63, 3454 PK de Meern, The Netherlands.* GERMANY: *PBV Proost & Brandt Verlagsauslieferung, Herzstrasse 1, 5000 Köln, Germany.* AUSTRALIA AND NEW ZEALAND: *Bookwise International, 1 Jeanes Street, Beverley, South Australia 5007.* THE FAR EAST AND JAPAN: *Japan Publications Trading Co., Ltd., 1–2–1, Sarugaku-cho, Chiyoda-ku, Tokyo 101.*

First edition: October 1987

LCCC No. 87–80495
ISBN 0–87040–691–4

Printed in Japan

Foreword

Takuo Fujita is an international authority on bone diseases and on nutrition, particularly as it affects the skeleton. Among his many activities around the world, he has been a member of the Board of Directors of the International Conferences on Calcium Regulating Hormones and that organization's host at its meeting in Kobe in 1983. Presently because of his special knowledge of calcium metabolism, he is a member of the Malnutrition Panel of the U.S.-Japan Cooperative Medical Sciences Program; the Japanese and American members of this panel meet regularly to advise on collaborative medical research between scientists in the two countries. From a number of years of early training in the United States, Dr. Fujita writes from a unique position of familiarity with the diets and life styles of both Western and Oriental peoples.

The information and recommendations in Dr. Fujita's book are based on sound physiological principles, including the fact that the human skeleton is ever-dynamic, changing in mass and strength under the influence of many factors. One of the most important of these factors is nutrition, calcium in particular being a major influence on the maintenance of bone health.

Dr. Fujita's book is highly recommended to all with a serious interest in becoming better informed—to their own benefit and that of their family—on calcium and its relationship to good health.

G. DONALD WHEDON, M.D.
Formerly, Director,
National Institute of Arthritis, Diabetes,
Digestive and Kidney Diseases,
National Institutes of Health,
Bethesda, Maryland, USA

Preface

Calcium is an important food component because it is an essential constituent of our body, especially the bones and teeth. Calcium deficiency therefore decreases the amount of calcium in these hard tissues, inevitably causing weakening. The actions of calcium in the human body, however, are by no means limited to the strengthening of hard tissues like bone and teeth. Calcium, indeed, is the most important nutrient for the maintenance of human life.

In the beginning was calcium. Life would never have been born without calcium. Unless the calcium concentration in blood is kept at an exactly constant level, the heart, brain and muscles cannot perform their functions. The human body is composed of billions of cells. Surrounded by a cell membrane sharply demarcating it from the blood or extracellular fluid, the interior of each cell maintains a fixed concentration of calcium only 1/10,000 the concentration of calcium outside the cell. Even a slight shift in such an immense difference in these calcium concentrations will result in emittance of a signal which could be responsible for cell division, movement, as well as generation of electrical current. Calcium, as a messenger, runs through each corner of our body to ensure that each cell can accomplish its part correctly.

Thus our human body performs its normal function by the work of calcium. When oral calcium intake is insufficient, the concentration of calcium in the blood falls. In order to supplement this, *parathyroid hormone* is secreted. Through its action even the important bones and teeth are dissolved to keep the concentration of calcium in blood at a vitally important constant level. Since an extremely large amount of calcium is deposited in the bone, dissolution of only a small portion of it causes an overflow of calcium through the blood vessels and brain. Since parathyroid hormone also brings calcium from outside to inside the cell, the amount of calcium in the cell increases, breaking the balance of calcium between inside and outside of the cell. This causes various disorders throughout the body, contributing to the development of many diseases. Thus an insufficient calcium intake in the diet causes an overflow of calcium within the soft tissue and within the cell. This is the *Calcium Paradox*. Such a Calcium Paradox may become the trigger for hypertension, *myocardial infarction*, *diabetes mellitus*, and even *senile dementia*.

Although calcium is thus one of the most important nutrients, and we are so profoundly deficient in calcium that an excess of dietary calcium is unexpected, an erroneous notion is widely held that "excess calcium intake is harmful, causing stones and arteriosclerosis." Excess calcium intake has been shown to cause kidney stones in only a very small number of people with unusual metabolic activity called *absorptive type of idiopathic hypercalciuria*. The majority of people are capable of adjusting the efficiency of intestinal absorption according to the requirement of the body for calcium by controlling the synthesis of active vitamin D and by other mechanisms. When excess calcium is coming into the intestine, the absorption efficiency is decreased. When calcium is deficient in food, our body tries its best to absorb it. Such adjustment is done so well that you need not worry about taking too much calcium by mouth unless you are a habitual kidney stone former because of idiopathic hypercalciuria. What you should be concerned about is the opposite occurrence, that is, not ingesting *enough* calcium, at which time calcium will be drawn from the bone to cause Calcium Paradox. Much of the misunderstanding about calcium may stem from an underestimation of the very special role of bone and the parathyroids in calcium metabolism.

Acknowledgment: The author is grateful to Miss Janet M. Lacey of University of North Carolina School of Public Health, who kindly reviewed the manuscript from an American readers viewpoint during her stay in Kobe for an epidemiological survey of osteoporosis.

Contents

1. Human Body and Calcium

1. Life Begins with Calcium

As far as we know, life is found only on earth, one of the smallest among millions of planets in the universe. At some point in the earth's evolution, vapors probably condensed, causing rain for an extended period of time, until the oceans were formed. The origin of life is still an enigma. Simple inorganic elements such as oxygen, hydrogen, carbon, and others form complex organic compounds. Protein is formed upon the addition of nitrogen. Something happens when lifeless materials such as protein become a self-productive organism, starting the eternal chain of reproduction. This is the mystery of life. When life was born on earth for the first time billions of years ago, great turmoil was probably taking place on the land and in the sea. Since life does not exist on the moon where there is no water, it may be safe to assume that the first life was produced near or in the water. It is thus widely accepted that life was born in the sea.

When we look at the sea, we may feel at ease and at home, possibly because our ancestors have come from the sea. Fish living in the sea at present are by no means our ancestors, but merely represent another branch of the evolutionary tree. Some unknown, simple, possibly unicellular primitive form of life in the sea may be the ancestor of all living creatures on earth.

Several reasons support the assumption that life originated in the sea. For example, the elements composing our body are quite similar to the constituents of seawater, but quite different from those of the soil. It is therefore more likely that materials making up our body have come from the sea and not from the earth. Calcium occupies the largest share among the inorganic components. Seawater contains abundant calcium. All living creatures coming from seawater are thus linked very closely to calcium.

When our distant ancestors left the sea to live on the land, their greatest problem was probably the lack of calcium in the air, compared to seawater. Every cell in their bodies was so used to the environment of calcium-rich seawater, that they could not live away from an environment like seawater. The blood circulating through our body is a substitute for seawater. Blood tastes salty like seawater as you might have noticed when brushing the teeth too hard. The blood is a form of seawater within our bodies. Like seawater, blood contains abundant calcium. Blood has been believed to be the source of life and the "water of life." Calcium is an essential part of blood, and the calcium concentration in blood is always kept relatively constant.

Fig. 1 Composition of the Soil, Seawater and Human Body. The ten most abundant elements listed in decreasing order. Except for phosphorus (P) in the human body and magnesium (Mg) in seawater, all other elements are found in both the human body and seawater. Soil, on the other hand, contains large amounts of other elements such as Silicon (Si), Aluminum (Al), Iron (Fe), and Titanium (Ti) different from seawater and the human body.

Decreasing Order of Abundance	Soil	Seawater	Human Body
1	O	H	H
2	Si	O	O
3	H	Na	C
4	Al	Cl	N
5	Na	Mg	Na
6	Ca	S	Ca
7	Fe	K	P
8	Mg	Ca	S
9	K	C	K
10	Ti	N	Cl

Oshima, Y. et al., *Cosmic Biology*, 1977, Kobun-sha

Fig. 2 Among the inorganic constituents, calcium and phosphorus are predominant.
Human Body Constituents

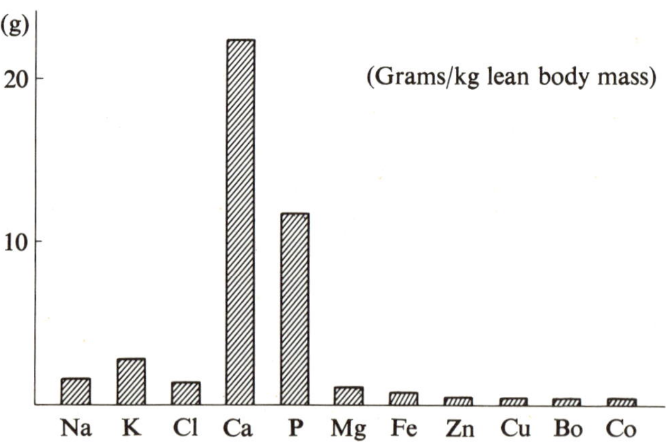

2. Consequences of Calcium Deficiency

What, then, is calcium doing in our body?

The largest part of calcium in our body, almost 99 percent, is found in the bone to make it hard enough to support the body. A much smaller amount of calcium is found in the blood, yet it is very important for controlling the action of the heart, brain function, hormone secretion, blood coagulation and other activities. The importance of calcium was first discovered in England about 100 years ago. Dr. Sidney Ringer found that in the absence of calcium, a frog's heart would stop. When the calcium level in the blood falls, muscle cramps and convulsions occur, and brain function decreases, causing irritability or even loss of consciousness. Heart action is also troubled, with the pulse becoming irregular and the contraction required to drive blood throughout the body becoming weaker. Finally, the heart may even stop beating. When blood calcium rises too high, again the heart does not beat properly and may even stop. Drowsiness and even unconsciousness may occur. Strict maintenance of the calcium level in blood is thus absolutely indispensable for life and for health. Even a minor change in blood calcium level causes disease, and a major change may cause death. Blood calcium level is thus called "one of the most precisely controlled biological constants."

When we are not taking enough calcium, does the calcium level in blood fall immediately?

If this should occur, it would be quite serious. But our body does not allow this to happen. Even if dietary sources of calcium are insufficient, the blood calcium level is maintained by a special mechanism.

When blood calcium level falls even a little, parathyroid hormone is immediately secreted to take calcium away from the bone, restoring the blood calcium level to normal. This is easy, because the bone contains more than 1,000 times the amount of calcium present in blood. Nevertheless this wonderful system is not without complications, if we depend upon it too often.

Fish living in seawater never become calcium-deficient because calcium-rich seawater is constantly coming in through the gills during respiration. All the fish have to do is to take as much calcium as necessary from this limitless supply. Fish have no parathyroid glands. *Amphibia*, the family of frogs living both in water and on land, have

Fig. 3 Creatures living in calcium-rich seawater (on the left) are surrounded by a rich calcium-containing environment and are constantly trying to expel thecalcium that is coming in (arrows directed to outside). Those living on land breathing air, (on the right), on the contrary, are in a calciumdeficient environment and try to keep as much calcium as possible (arrows directed towards inside).

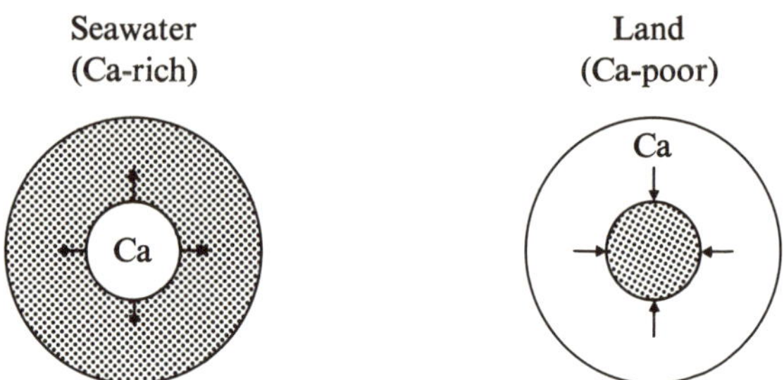

the beginnings of simple parathyroids, and all living creatures on land have parathyroids, probably because they need them to compensate for calcium deficiency. Parathyroid glands are like keys, which can take calcium out of the deposit in the bone when calcium intake is insufficient, and when a danger of a decreased serum calcium level is imminent. Parathyroid hormone, secreted from the parathyroid glands located near the thyroid gland, is one of the three calcium regulating hormones.

Calcium is like money deposited in a bank called "bone." Just as a person takes cash out of the bank with a cashcard whenever necessary, parathyroid hormone can take calcium out of the bone when dietary calcium is not enough to keep the blood calcium level constant.

Even if you have a huge deposit in the bank, it will decrease if you keep drawing from it. The calcium deposit in the bone also decreases if calcium is constantly being withdrawn. In consequence, not only does the bone becomes weak and susceptible to *osteoporosis*, but also other tissues may suffer damage. This can be a serious menace to our health. Deficiency of calcium in food clearly should not simply be compensated for by drawing calcium out of the bone.

It is difficult to take out exactly the amount of calcium required by the body, unlike the computerized withdrawal system in the bank. In order to meet the emergency of calcium deficiency threatening life, more calcium than necessary is frequently released from the bone. Since the blood calcium level has to be kept constant at all times, the excess calcium cannot stay in blood. The overflowing calcium cannot be readily returned to the bone or excreted in urine. Some of the calcium enters the soft tissues such as blood vessels and brain, or even inside the cell. Such "soft tissue" calcium deposition threatens health and may even cause fatal diseases. Another reason for such an overflow of calcium is the vast difference in calcium content between the bone and soft tissues. Bone contains about 1 kg or 1,000,000 mg calcium. Only 200 mg of calcium or 1/5,000 the amount in bone is contained in our entire 5 liters of circulating blood. When only a fraction of calcium in the bone is released, it readily causes a flood of calcium into the blood and soft tissues.

Dietary calcium is different from calcium withdrawn from bone. Even if too much calcium is taken by mouth, the gut absorbs only amount necessary for the body at that time. Strictly speaking, the digestive tract from the mouth to the anus is still outside of the body. The actual entrance into the body is not the mouth but the gut wall. A small proportion of calcium is absorbed automatically or passively. A larger portion is absorbed actively with the help of vitamin D, only when calcium is necessary for the body. When too much calcium is found in the food, active absorption decreases and the total calcium absorption from the gut is markedly reduced. Such selection by the gut and adaptation to intake is more effective for calcium than other substances, probably because calcium is so important. In some people, however, such a delicate control of calcium absorption by the gut does not work effectively. The gut cannot adapt to the amount of calcium in the body as accurately as it should, so that too much calcium may be absorbed even when it is not necessary for the body. Since excess calcium entering the body must be excreted by the kidney, stones are frequently formed in people with so-called idiopathic hypercalciuria. Naturally, for such people, taking too much calcium is dangerous and should be avoided, beause it would be just like injecting calcium directly into the blood. Normally, however, in most people, intestinal calcium absorption is controlled according to the intake. In anyone, calcium directly injected into the blood could damage the tissue, but calcium taken by mouth is carefully measured and controlled and only a small, necessary amount is absorbed from the gut. Calcium coming through the gut is controlled and harmless, probably because the amount is much smaller than the amount of calcium released from bone.

When the calcium intake is insufficient calcium comes out of the bone and the bone starts to become weaker. The calcium which cannot stay in the blood overflows into soft tissues, and into cells.

Calcium deposited in the blood vessel wall can lead to *arteriosclerosis* and hypertension. Calcium deposition in the brain contributes to deterioration of brain function and dementia. Calcium deposition in the coronary arteries and heart muscle may cause myocardial infarction and decreased strength of heart muscle.

What happens when *sufficient* calcium is taken by mouth? Normal blood calcium level is easily maintained and the bone has enough stored calcium to be strong; parathyroid hormone is secreted normally and no excess calcium comes out of the bone. In addition, the blood vessels do not suffer from a flooding of calcium from the bone, and they stay useful longer.

When calcium is deficient in food, calcium excess occurs in the tissues and cells. This is Calcium Paradox and is unique to calcium because of its storage in the bone and hormonal control via the parathyroid gland. The term "Calcium Paradox" is sometimes used in another sense. When heart muscle cells made deficient in calcium are then placed in a solution containing calcium, the cells are destroyed by calcium. This is the Calcium Paradox in physiology. In order to distinguish it from the phenomenon of calcium deposition in the blood vessels and brain in calcium deficiency, we might call the latter the *Calcium Paradox of Aging*.

Too much dietary calcium does not normally cause calcium stones or calcium deposition in the blood vessels leading to arteriosclerosis. On the contrary, it may prevent the really damaging calcium outflow from the bone. Calcium deficiency, on the other hand, prompts calcium deposition in tissues and cells (Fujita, 1986).

Fig. 4 Aging and calcium. The first triangle shows an increase in the degree of calcium deficiency. The second triangle indicates an age-bound increase of parathyroid hormone secretion in response to calcium deficiency. The third triangle representing bone calcium shows a decrease with advancing age, leading to osteoporosis. The fourth triangle representing soft tissue calcium increases with age, contributing to the development of hypertension, arteriosclerosis and senile dementia.

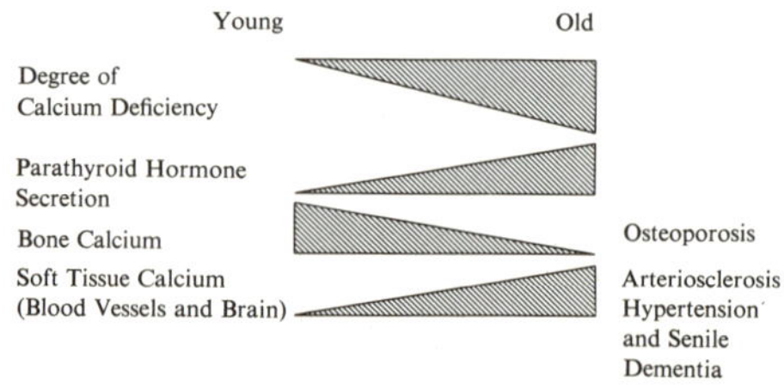

3. Actions of Calcium

Our body is made up of about 60 trillion cells. When we look at our body in the mirror, we see head, chest, arms and legs. Since the cells are too small for our eyes, we cannot see the cells making up our body. However, when we look at muscle, liver and brain specimens under the microscope, we can easily identify individual cells. Cells are everywhere in the human body.

For instance, blood is not just a red-colored fluid. Billions of red and white cells are floating in the plasma. The liver consists of *hepatocytes* and other cells, neatly bound and piled up, irrigated by blood and bile. The brain consists of about 14 billion nerve cells interspersed with *glia cells.* Nerve cells characteristically give out long nerve fibers. These are protrusions of the cell body and part of the cells.

Many organs of different shape and function are thus found in our body. All these organs are collections of cells assembled by connective tissue or intercellular substance. Blood is the most simple form of tissue, consisting of cells floating in fluid. Diseases of our body are diseases of individual cells, and our health represents the total health of all individual cells.

What is the healthy environment for each cell? Each cell is surrounded by a cell membrane (*plasma membrane*), to clearly distinguish inside from outside. It is the responsibility of the cell membrane to draw a sharp borderline between the inside and outside of the cell, preventing the free inflow of outside substances and the outflow of inside substances. In case the cell membrane becomes idle, the vitality of a cell is immediately lost.

Concentrations of various substances differ between inside and outside the cell. For example, sodium is more abundant outside the cell but cannot easily enter the cell. On the contrary, more potassium and magnesium are found inside the cell than outside, and they cannot readily go out. The inside-outside differences of these elements are only several to 100-fold at most.

Calcium is a quite exceptional substance in this regard. A constant concentration of calcium is always found in blood or extracellular fluid (about 10 mg/dl). Inside the cell, however, only 1/10,000 of this level is found. The concentration difference between inside and outside the cell is thus as high as 10,000 times, at least 100 times higher than the highest outside-to-inside differences of other substances.

Whenever a cell loses its vigor, becomes older, damaged or diseased, calcium inside the cell increases. For example, when calcium in the cell increases from 1 unit to 5 units, and calcium outside the cell remains at 10,000 units, the outside-to-inside ratio of calcium promptly decreases to 2,000: 1. When calcium increases to 20 units, the ratio decreases to 500: 1.

When the inside and outside calcium concentrations become exactly the same, the cell membrane is of no use and the cell is dead. The most important function of the cell membrane is to protect the cell by preventing unnecessary calcium from entering the cell. When the cell becomes sick and the action of the cell membrane becomes weak, it is no longer able to prevent the undesirable entry of calcium into the cell. Entrance of calcium further weakens the cell function in a vicious cycle finally leading to cell death.

How is the cell membrane able to prevent undesirable calcium from entering the cell?

The cell membrane is like the immigration control at the airport. Calcium units entering the cell from outside are checked one by one, and only the desirable ones are permitted to enter. Unless you have a proper passport, you cannot pass through the immigration control. Parathyroid hormone is like a passport authorizing calcium to enter the cell. Calcium also enters the cells, when the electrical potential changes between the inside and outside of the cell membrane. When this occurs in a nerve cell, excitation is said to occur. Such electric current is then transmitted along the nerve fiber. This calcium gate may be opened when certain hormones such as parathyroid hormone arrive at the cell membrane and react with receptors. This unlocks the "gate" for calcium to enter, much like a key fits into a keyhole to open the door.

In order to take unnecessary calcium out of the cell, a mechanism called the *calcium pump* is in action. It is like pumping water out of a boat floating on water. The calcium pump in the cell membrane is always busy because it has to keep the calcium inside the cell at a level much lower than that outside the cell.

How does calcium enter the cell when calcium in the food is insufficient? When the blood calcium level decreases even slightly, parathyroid hormone is secreted in response to this signal. Since parathyroid hormone is like a passport to authorize calcium to enter the cell, calcium enters the cell easily with the help of the parathyroid hormone. When parathyroid hormone is injected, blood calcium at first slightly decreases, probably because of the entrance of calcium into cells, but soon much more calcium comes out of the bone and blood calcium rises again. Other hormones may also act on a receptor to open the calcium gate, but parathyroid hormone is most sensitive to the decrease of serum calcium and calcium deficiency in the body.

The vast concentration difference between inside and outside the cell is maintained at all costs to enable calcium to accomplish its work as a messenger.

Calcium coming inside the cell, through the strict barrier of the cell membrane, is transmitting a signal from outside, and is a precious guest on account of its rarity. When the amount of calcium inside the cell increases, calcium entering the cell from outside is no longer rare, and it is no longer able to accomplish the work as a messenger. Calcium should be brought out of the cell to keep the concentration difference always high.

The cell is like a hotel for calcium. It admits only a limited number of guests. Only guests with reservations are permitted to stay, thus "Many are called, but few are chosen." Inside the cell hotel, calcium

Fig. 5 Calcium distribution shows a marked difference in concentration between outside and inside of a living organism and among various compartments within the organism. For creatures living in seawater, (shown on the right), calcium content of the seawater (spots) is much higher than the calcium content in blood (blank). A huge amount of calcium is stored in bone (shaded area). Each cell of a living organism has a similar calcium distribution. Cytosolic calcium concentration is much lower than the calcium concentration in blood, and mitochondria in the cells store a large amount of calcium, like the bone in the organism, as shown on the left.

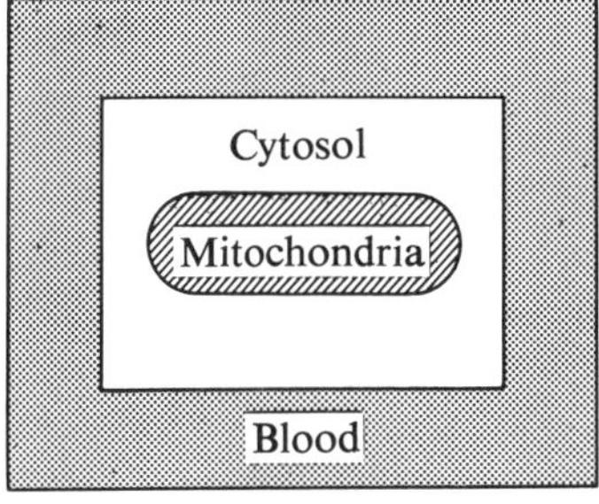

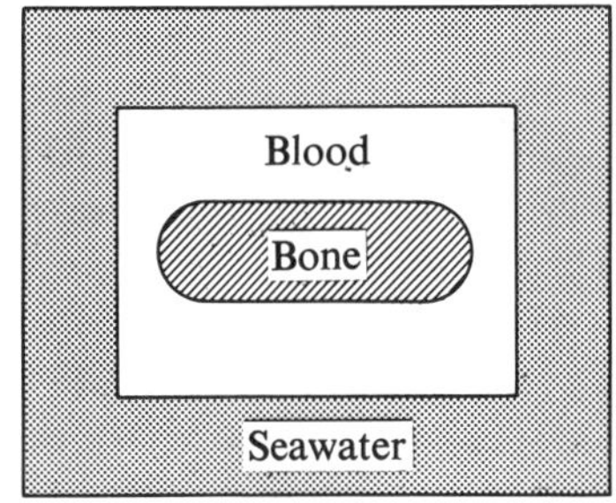

Vast Intercompartmental Differences
in Calcium Concentration

guests stay in various rooms or compartments such as *mitochondria* and *microsomes*. Calcium binding proteins in the cell, such as *calmodulin*, make sure each calcium stays where it belongs, without overflowing from its cellular compartments. All these systems are designed to minimize the calcium concentration in the intracellular fluid. The high inside-outside difference of calcium is thus maintained to keep the shape and function of the cell membrane and to insure the messenger activity of calcium. The messenger system within the cell is extremely complex. Calcium is not the only messenger, but is by far the most important, playing a key role in coordinating various other entire messenger systems.

Calcium exerts three major actions on the cell. First, it keeps the cell alive, playing important roles in cell proliferation or division, and differentiation or gradual change of shape and function. When calcium enters the cell from outside, the *cytoskeleton*, consisting of *microtubules* and *microfilaments*, is activated and the cell starts to move. Calcium is necessary for amoebic movement of white blood cells and *phagocytosis*, the ingestion of foreign bodies such as bacteria. Unless calcium enters the cell, the cell stays dormant and immobile.

Second, calcium causes excitation or electrical activity of the cell. For example, within the eye, retinal cells respond to the light with the entry of calcium into the cell. A substance called *rhodopsin* is transformed on exposure to light and allows calcium to enter the cell. An electrical current is generated and this is transmitted via the nerve fibers to the cerebral cortex, where it is perceived as light. Although the electricity generated by a cell itself is small, it is quite potent if one considers the thickness of the cell membrane. When we calculate the electricity generated by the cell relative to the cell membrane, assuming that the cell membrane is as thick, proportionately, as the wall of a building, the electricity corresponds to 30,000 volts! This enormous electricity generated by the cell membrane is one of the mysteries of life. Calcium produces such electricity, which is transmitted all around the human body forming a messenger network system. Calcium is thus extremely important in message transmission.

Third, the secretion of hormones and other materials from the cell also requires calcium. Hormones are messengers transmitting the information among cells, mostly through the blood stream. For example, when sugar enters the blood after meals, calcium enters the *B-cells* of the *Langerhans islet* of the pancreas as a signal to secrete insulin in order to metabolize the sugar for energy. Calcium is also necessary to transport the hormone into and out of the cell. Imbalance of calcium between inside and outside of the cell makes adequate insulin secretion difficult. Calcium is required for the secretion of the growth hormone to assure normal child growth, *prolactin* to stimulate lactation, *adrenocortical hormone* to endure stress, and all other hormones.

The cell is also like a factory. It makes many things in addition to hormones. In order to digest food, stomach cells must make gastric juice and pancreatic cells pancreatic juice. Calcium is always required for such production, such as many kinds of white cells and *lymphocytes*. These cells are important for the immune function because they destroy bacteria and other harmful foreign materials with antibodies that act like missiles. *Cytokines*, acting as messengers between these cells, are also produced in response to a calcium message.

Thus calcium is indispensable for the function of each cell and the mutual network of connections between cells.

4. Calcium Regulating Hormones: Three Hormones that Control Calcium Metabolism

Calcium which enters the body as food is absorbed from the intestine and enters the bloodstream. In order to keep the calcium level of the blood constant, calcium regulating hormones are watching the movement of calcium all the time. Parathyroid hormone, *calcitonin*, and the active form of vitamin D (1, 25-dihydroxy-vitamin D) are called *calcium-regulating hormones*. Vitamin D occurs naturally in fatty fish, eggs, liver and butter. Milk and margarine are fortified with vitamin D in the U.S. While 2.5 μg (100 U) vitamin D is sufficient to prevent rickets, 10 μg is desirable to secure good calcium absorption in children (RDA). This is reduced to 7.5 μg between ages of 19 to 20 and 5 μg after 22 years. Vitamin D contained in food or synthesized in the skin by the help of ultraviolet ray is not active, in the sense that it cannot stimulate intestinal calcium absorption or bone resorption unless it undergoes some chemical changes. The liver adds one hydroxyl group to produce 25-OH-vitamin D, and then the kidney adds the second hydroxyl group at the 1α position to produce $1\alpha 25(OH)_2$ vitamin D, which is sometimes called *active vitamin D*. Since addition of the first OH at 25 is done rather freely without restrictions, 1α(OH) vitamin D, a synthetic product used as a drug, is also called active vitamin D. They control calcium, and calcium controls them. Without the action of these hormones, blood calcium would be changing all the time.

The blood calcium level changes very little regardless of the amount of calcium in food. When dietary calcium intake is insufficient, the calcium level in blood may fall slightly, but this is immediately restored to the normal level by the parathyroid hormone, as long as the parathyroid glands are functioning normally. When the parathyroid glands are incapable of producing the necessary amount of parathyroid hormone, serum calcium falls. On the other hand, too much parathyroid hormone raises blood calcium. A deficiency of vitamin D also decreases serum calcium level.

As their name indicates, the *parathyroid glands* are two pairs of rice-sized glands located near the thyroid. Until about 100 years ago, nothing was known about the important actions of these small endocrine glands and how they control calcium movement within the body and play an important role in taking calcium out of the safe deposit called bone. Ivar Victor Sandström of Sweden was one of the first to describe these glands as endocrine glands independent from the thyroid, but many years passed before anybody paid attention to his report.

In the beginning of this century, research workers removed the thyroid glands from dogs to study the function of the thyroid glands. When the thyroids were removed, the dogs went into convulsion and died, and the experiments ended in failure. It then occurred to the researchers that some strange glands called parathyroids were reported to be present near the thyroid gland. After a careful search, they found out that during their experiment, the parathyroids had been removed along with the thyroids. In a second attempt, they removed the thyroids and left the parathyroids untouched, and the dogs remained quite healthy, without going into convulsions. On removal of the parathyroid glands alone, the animals suffered from convulsion and death. The blood calcium levels were also very low in these animals. This experiment connected the parathyroid gland with calcium for the first time. We have learned from these experiments that in dogs the parathyroids are more important for immediate survival than the thyroids. Parathyroid hormone was thereafter extracted from the parathyroid glands. Injection of parathyroid hormone into dogs raised blood calcium levels. Dogs were therefore used for many years to measure the potency of parathyroid hormone preparations.

Parathyroid hormone raises the blood calcium level through several mechanisms. First, it takes calcium from the bone and releases it into the blood. Second, it decreases urinary calcium excretion to prevent calcium loss. Third, it stimulates the kidney to produce the active form of vitamin D, which in turn increases calcium absorption from the gut. Thus the effects of parathyroid hormone may be summarized as the increase of the calcium transfer from bone, kidney and gut to the blood. As more parathyroid hormone is secreted, the blood calcium level will also rise.

In *primary hyperparathyroidism*, a disease in which too much parathyroid hormone is produced, the blood calcium level is always high. In *hypoparathyroidism*, a disease in which too little or no parathyroid hormone is produced, the blood calcium level is generally low.

Parathyroid hormone and the active form of vitamin D work very closely together. Parathyroid hormone produces the active form of vitamin D in the kidney, just as parents give birth to children. For drawing calcium from bone, they work together like brothers and sisters. In the absence of active vitamin D, parathyroid hormone cannot take calcium from the bones as easily. Active vitamin D also requires parathyroid hormone for its best action. The blood calcium level is maintained at a proper and constant level as a result of cooperation among the calcium regulating hormones. Parathyroid hormone secretion rises with age along with the progress of calcium deficiency (Wiske et al, 1979). While the best signal for parathyroid hormone secretion is a fall of the serum calcium level, parathyroid hormone secretion is also stimulated when the concentration of active vitamin D in the blood is low, even in the face of a normal serum calcium level.

Vitamins and hormones are usually entirely different substances. When active vitamin D or $1,25(OH)_2$ vitamin D is counted as one of the calcium regulating hormones, you may be afraid that there has been a mix-up between vitamins and hormones. A *vitamin* is a nutrient found in foods. When a vitamin enters our body, it helps the metabolic process without itself being changed. A *hormone*, on the other hand, is a product of the human body. Raw materials for hormonal production may come from food, as in the case of cholesterol which is the starting material for steroid hormone synthesis. Vitamin D is such a raw material contained in various foods, but vitamin D by itself must be transformed within the body to be active. Most of the vitamin D in our body is made from 7-dehydrocholesterol by the energy of sunshine (ultraviolet ray).

Vitamin D is then converted to 25 (OH) vitamin D in the liver and brought to the kidney. Here in the kidney 25(OH) vitamin D can be changed to 1,25(OH)$_2$ vitamin D, the final active form of vitamin D. This conversion occurs when serum calcium is low and parathyroid hormone secretion is stimulated or when serum phosphorus is low; however, 25(OH) vitamin D is converted to 24,25(OH)$_2$ vitamin D when serum calcium is high and parathyroid hormone secretion is inhibited. While 1,25(OH)$_2$ vitamin D is a very potent substance, the action of 24,25(OH)$_2$ vitamin D has not yet been definitely established. It may have some effect, but it is much less active than 1,25(OH)$_2$ vitamin D for stimulating intestinal calcium absorption. Like a factory

Fig. 6 Conversion of 7-dehydrocholesterol to vitamin D (cholecalciferol) by sunlight (ultraviolet ray).

Ultraviolet rays in the sunlight cut the ring of 7-dehydrocholesterol, a precurser of vitamin D, in the skin, like a pair of scissors. Vitamin D is thus synthesized in the skin.

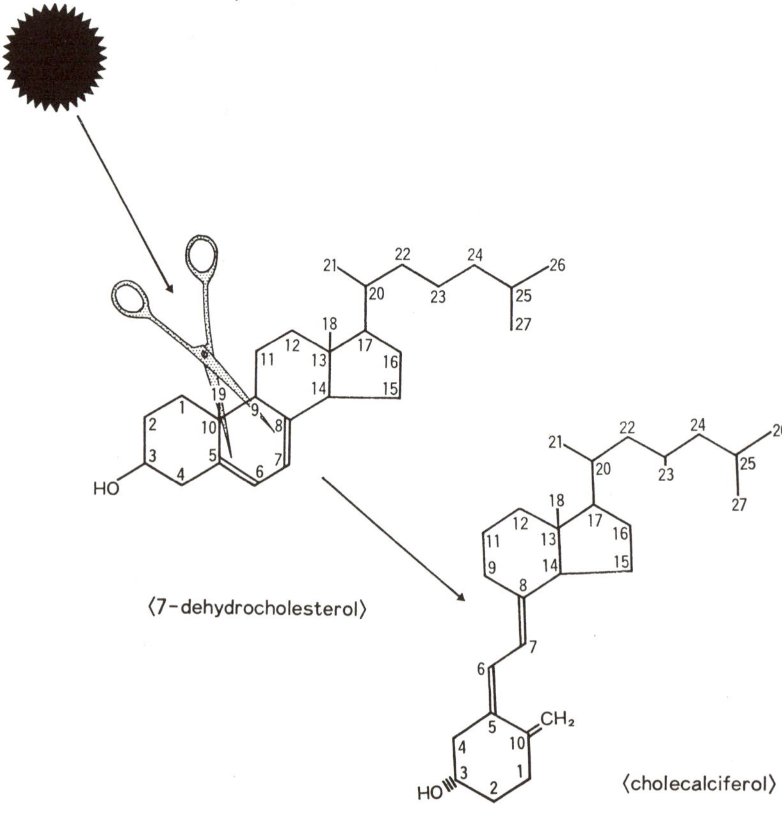

⟨7-dehydrocholesterol⟩

⟨cholecalciferol⟩

producing machines, the kidney tries to make what the body really needs. Since 1,25$(OH)_2$ vitamin D is actually a product of the human body, it is in fact a hormone and no longer a vitamin.

Why doesn't blood calcium rise endlessly while these two powerful hormones strive to make serum calcium higher? The human body is

Fig. 7 After conversion from 7-dehydrocholesterol in the skin by the help of ultraviolet light, vitamin D is converted to 25-(OH) vitamin D in the liver and circulates in the blood. In the kidney, it is converted to 1,25$(OH)_2$ vitamin D when serum calcium is low and the body needs more calcium. When serum calcium is high and the body needs no more calcium it is converted to 24,25$(OH)_2$ vitamin D, a relatively inactive compound.

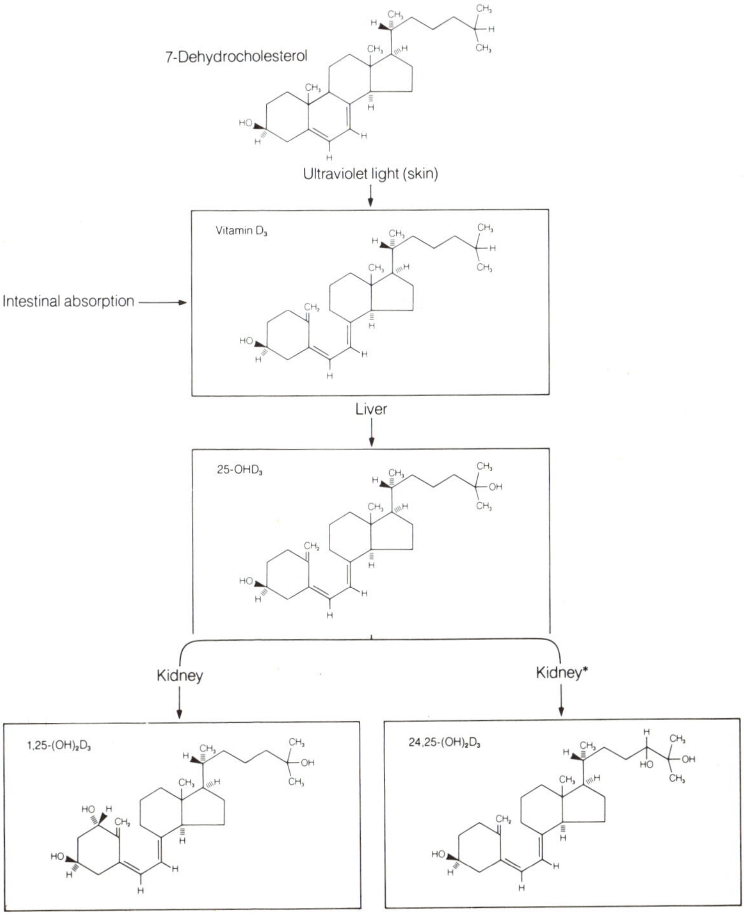

7-Dehydrocholesterol

Ultraviolet light (skin)

Intestinal absorption

Vitamin D_3

Liver

25-OHD_3

Kidney

Kidney*

1,25-$(OH)_2D_3$

24,25-$(OH)_2D_3$

so well designed that some hormones are always checked by others in order to maintain a stable balance. Calcitonin, produced by the thyroid gland, acts to increase calcium flow from the blood to the bone, thus antagonizing the action of parathyroid hormone and the active form of vitamin D. When blood calcium is too high, no parathyroid hormone is produced. Rather calcitonin is produced, in order to prevent a dangerous rise of blood calcium.

In view of these actions of parathyroid hormone and calcitonin, it is quite understandable that fish living in calcium-rich seawater produce no parathyroid hormone, but instead produce abundant calcitonin, secreted from the *ultimobranchial glands.*

In all the animals living on land, parathyroid glands are found, and enough parathyroid hormone is produced, but not much calcitonin is secreted, especially as the land animals become older. In a state of calcium deficiency, they probably would not require calcitonin. It is of interest that women produce less calcitonin than men, especially as they grow older, when many of them develop osteoporosis. Calcium deficiency is probably so severe in these women that no calcitonin is required for the maintenance of blood calcium (Deftos, 1981).

2. Human Life Cycle and Calcium

1. Pregnancy and Calcium

The beginning of an individual's life is the meeting of the *oocyte* with the *spermatozoon* during fertilization. Like all moving cells, the small spermatozoon requires calcium for its active movement, to find and reach the large oocyte quietly sleeping in the ovary. For the maturation and movement of spermatozoon, an adequate calcium intake and a proper calcium gradient across the cell membrane is an absolute requirement. If calcium intake is deficient and such a gradient across the spermatozoon cell membrane is blunted, no active movement takes place, so that it cannot find the oocyte. For simple cells and primitive organisms, movement is initiated by a change of shape. Muscle contraction is a good example of this. The entrance of calcium into the muscle cell and its binding with some protein causes its contraction. These proteins are called *contractile proteins.* Contractile proteins are found in all cells which change shape and/or move. *Flagella* or *cilia*, which propel the primitive organisms such as bacteria, also consist of calcium-binding contractile proteins. In the presence of abundant calcium, these organisms move around actively. The spermatozoon also has a flagellum which can move via calcium-dependent contractile proteins.

When the spermatozoon finally reaches the oocyte, it again uses calcium as the signal. As soon as the spermatozoon touches the oocyte, calcium flows inside the oocyte. Evidently, the oocyte should also keep a vast inside/outside difference of calcium concentration to keep itself as active as the spermatozoon. Otherwise, the calcium signal provided by the attachment of spermatozoon would not be properly received.

As soon as calcium flows into the oocyte, endless cell division and differentiation are started until a complete human being is formed. In lower animals, it is even possible to start the cell division by injecting calcium into the oocyte without spermatozoa (Campbell, 1983). Cell division and differentiation may also be started this way, although the organism thus produced would have only one-half the usual number of chromosomes. This is called *parthenogenesis* or *virgin birth.* Calcium itself could even take the place of spermatozoa in signaling the oocyte. Calcium, indeed, is the source of life!

When the husband or wife is not taking enough calcium, it is possible that the movement of spermatozoa may be less active than it should, or the oocyte may not be completely ready, so that sterility might result. In order to avoid this, sufficient calcium intake is recommended.

After successful conception, pregnancy has to continue for almost 10 lunar months. It is not an easy job to carry the baby for this long a period. Morning sickness comes in the first part of pregnancy. *Toxemia of pregnancy* sometimes leads to kidney disease which may persist throughout life. The most severe complication of pregnancy is *eclampsia*, with hypertension, renal failure, convulsions, and even loss of consciousness. This could even be fatal, or may have long-term negative effects.

It was recently suggested that such dreadful complications of pregnancy may be the result of calcium deficiency, because calcium deficient mothers frequently suffer from such complications. It is very easy for a pregnant woman to become deficient in calcium. In order for the fetus's bone to form, the mother has to provide a large amount of calcium. Ordinary food does not contain enough calcium to carry the mother and baby through the pregnancy. When sufficient calcium is not available to maintain a normal serum calcium level, calcium comes out of the bone and is deposited in the blood vessels and kidney. This is an example of the Calcium Paradox. The blood vessels contract when too much calcium comes inside the smooth muscle cells of the blood vessel wall. This may result in a rise in blood pressure, or hypertension, and a decreased blood supply to vital organs such as the brain and kidney. Convulsions and renal failure are the consequences. Sufficient dietary calcium plus necessary calcium supplements would help to prevent dreadful complications of pregnancy such as eclampsia.

Everybody knows that a pregnant woman should take more calcium as food for her baby, but the question is, how much? All the calcium in the baby's bones must come from the mother. If a baby weighs 3,500 grams at birth, then at least 30 grams or 30,000 mg of calcium should be present in the skeleton of this baby.

How could the mother provide this amount of calcium? Since only about 700 mg of calcium per day is contained in ordinary food, and less than half of this amount, about 300 mg, is absorbed from the gut and finally utilized, the mother has to sacrifice most of the calcium from her daily dietary intake to make the bones of the baby for as many as 100 days. This covers a good part of the entire pregnancy period. The need for taking extra calcium is quite evident, because the mother herself cannot get along without dietary calcium, unless she sacrifices her bones to maintain blood calcium at the normal level. As a simple rule of thumb, the mother should take 1.5 times as much calcium during pregnancy as before, 1,200 mg compared to 800 mg, the requirement for non-pregnant women. Otherwise she will rapidly lose calcium from her bone. This would be detrimental, especially since calcium is indispensable for the maintenance of a firm pelvis, needed for safe delivery of the baby. Postpregnancy osteoporosis, the consequence of losing much calcium from the bone during pregnancy, is fortunately transient in most cases, because after delivery the mother is once again able to use dietary calcium for her own physiological needs.

Even if the baby draws all necessary nutrients from the mother and grows well in the womb, the baby cannot pass through a deformed pelvis. The normal pelvis is so well designed to allow the baby to pass through, that even a minor change in the shape of the pelvis presents an obstacle to the baby's head, which is usually the first part to pass through the pelvis. In order to maintain strong bone, sufficient calcium should be supplied. In case the mother loses too much calcium, her pelvis inevitably becomes too weak to resist the gravity and weight of the muscles. A narrow pelvis could result which may interfere with the safe passage of the baby's head.

In addition to calcium, a female hormone, *estrogen*, also plays a role in protecting the pelvis. Estrogen is secreted in much larger amount than usual during pregnancy, to help the uterus and ovaries to grow in preparation for fertilization and also to permit safe continuation of the pregnancy. At the same time estrogen protects bone and prevents too much calcium being withdrawn from it.

During pregnancy, calcium frequently becomes deficient. Sometimes the bone is dissolved to take calcium out to meet this need. Estrogen reduces the amount of calcium taken out of the bone. It also increases calcium absorption from the gut by stimulating the production of active vitamin D. By the action of estrogen, the body makes the best use of calcium contained in the diet. When a woman reaches menopause, menstruation ceases because of a sudden decrease of estrogen secretion. At this time, the bone also loses its good friend, estrogen, which had previously protected it from losing calcium. Consequently, the bone starts to lose calcium rapidly and such bone loss could cause *postmenopausal osteoporosis*, the most widespread disease among elderly women. Much more calcium is required to keep a calcium balance in postmenopausal women than in younger ones, due to the loss of estrogen which has previously done a good job in economizing calcium.

During childbirth, the so-called labor pains are caused by regular contractions of the uterus that push the baby and placenta out. This is the most important natural process of delivery. Muscles in the wall of the uterus are smooth muscles which cannot be controlled at will but spontaneously contract according to the body's needs, unlike the muscles of the arms and legs or skeletal muscles. The most important factor controlling the contraction of both smooth muscle and skeletal muscle is calcium. Muscle fibers are rearranged when calcium comes in, resulting in a remarkable shortening. This is muscle contraction which generates a tremendous strength.

Strong and regular contractions of the uterine muscles are induced only in the presence of the proper inside/outside difference of calcium across the cell membrane. For such strong uterine contractions, adequate calcium intake is naturally required in order to keep the steep concentration gradient of calcium across the cell membrane. When calcium intake is insufficient, such a balance is lost, and the occurrence of regular contractions is no longer expected. Safe delivery of the baby is not possible unless regular uterine contractions continue. Calcium is the source of life to give birth to the baby.

After the baby is safely delivered, the major work for the mother becomes feeding the baby with milk. Mother's milk contains smaller amounts of calcium than cow's milk or artificial milk, but the calcium in mother's milk is aborbed very efficiently from the gut of the baby. In premature and underweight babies, mothers milk alone cannot provide enough calcium, but healthy babies delivered at term do best with mother's milk.

In order to produce good mother's milk, the mother naturally has to consume enough calcium. At least 1.5 times the ordinary requirement is said to be needed for the mother breast-feeding her baby.

2. Child Growth and Calcium

(1) Body growth and calcium

When General Douglas MacArthur came to Japan in 1945, as the Supreme Commander of the Occupying Forces, what he saw were short, stooped and silent people, in contrast to the tall, handsome and merry GIs. That was why he thought Japanese people were as "immature as 12-year-old children." Strangely enough, *Niseis* or Americans of Japanese extraction among the GIs were as tall as their Western colleagues, completely unlike their Japanese grandparents. Evidently what they ate, and not what they had inherited from their ancestors, was more important in determining stature. Forty years have passed, and the Japanese people have been transformed. The mean height of the high school students increased seven inches over these years. Nobody says Japanese people are short any more. They started to drink milk and eat dairy products just like Americans and grew taller. They have even started to talk a little louder. These are some of the results of an increase of calcium intake.

Fig. 8 Calcium intake is gradually increasing, but is still low in Japan.

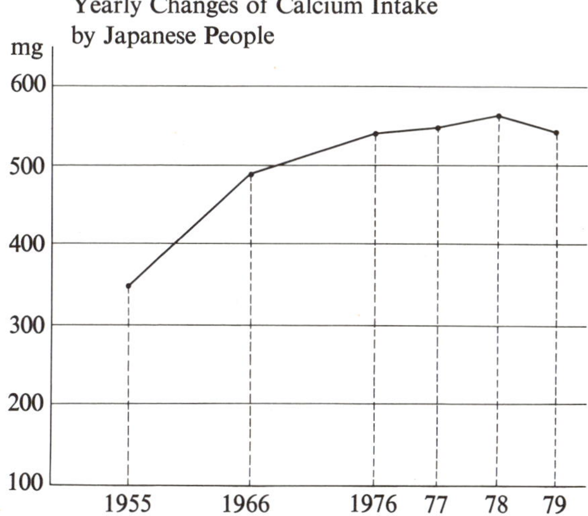

Traditional Japanese food contains only a small amount of calcium, being almost entirely free of milk and cheese, the main sources of calcium in Western diets. Although Japan has been a small overpopulated island since ancient times, the people have worked hard to produce rice and vegetables and catch fish to avoid protein-calorie malnutrition. Since the Tokugawa government prohibited the eating of any kind of animal meat, forcing the Japanese to depend on fish and soybeans for their protein intake, dairy products did not appear on the dinner table of Japanese for hundreds of years. This calcium lack was probably the major factor responsible for the short stature of Japanese. In an epidemiological survey in Wakayama Prefecture, central Japan, people living in the mountain region where calcium intake was low and sunshine was poor, were shorter than those living in the seacoast district, where calcium intake was high and sunshine abundant (Fujita et al, 1978).

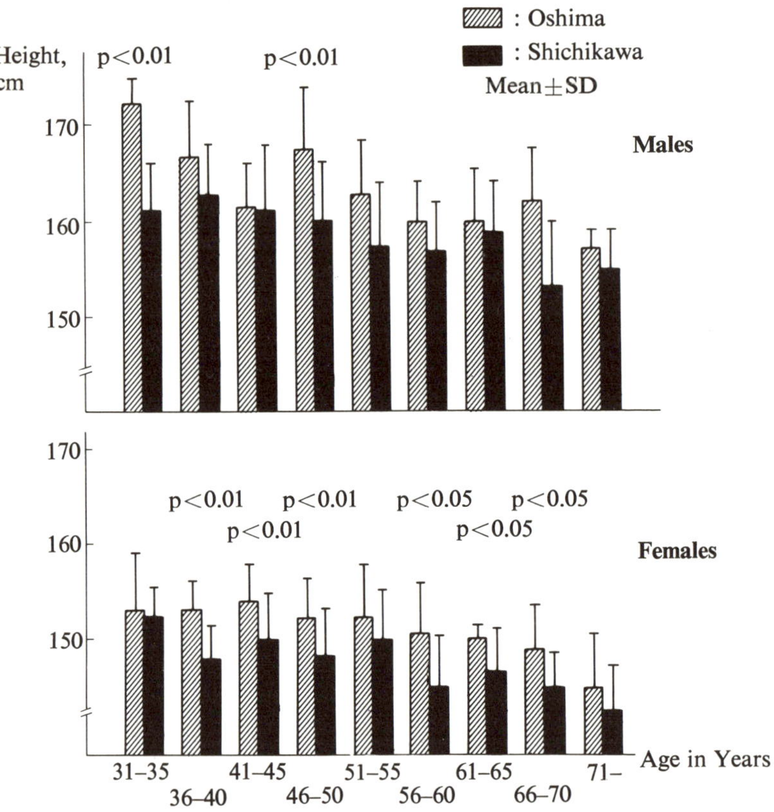

Fig. 9 Height and calcium intake.
The hatched bars represent the height in the seacoast Oshima district, with abundant sunshine all year round and adequate calcium intake. The white bars represent the height in the Shichikawa district high in the mountains where calcium intake is poor and exposure to sunshine is less.

In Japanese homes, sons are much taller than their fathers and daughters are far taller than their mothers, almost invariably. Parents are happy to look up to their tall sons and daughters. This is the symbol of the post-war development of the Japanese economy, or the result of "calcium shock."

Children who are short because of calcium deficiency may become taller in response to calcium supplementation, administered during the growth period. However, calcium deficiency is not the only cause of short stature. Not all short children suffer from a lack of calcium.

Congenital bone disease is difficult to treat. Deficiency of growth hormone or thyroid hormone should naturally be treated by supplementation of these hormones. Calcium alone might not be effective for short stature caused by these factors. Even the actions of growth hormone and thyroid hormone, however, require calcium and adequate calcium gradient across cell membranes. Sufficient calcium intake is therefore necessary to make sure these hormones accomplish their purposes and to supply the material for bone growth. The hereditary factor providing the background or the outline of growth speed might be modified but probably cannot be changed completely.

It is nevertheless important to check for possible calcium deficiency in children who are not growing as they should.

Fig. 10 Incidence of fracture in children and youth in Japan (%). The incidence is increasing at all age levels of children.

	1970	1972	1974	1976	1978	1981	1983
Kindergarten	0.17	0.21	0.19	0.18	0.20	0.23	0.24
Primary School	0.53	0.70	0.64	0.66	0.74	0.75	0.81
Junior High School	1.07	1.24	1.14	1.25	1.36	1.42	1.48
Senior High School	0.64	0.61	0.62	0.69	0.79	0.73	0.80

In recent years, fractures in children have reportedly been increasing in Japan. Since calcium deficiency takes calcium away from the bones, making them weaker, such a deficiency may represent the major factor in childhood fractures. Although calcium intake has been increasing in Japan, this might not have been enough to decrease fracture incidences. Simultaneously, increased intakes of protein and phosphorus may have antagonized the beneficial effects of the increased calcium intake. Insufficient exercise might also be responsible because in addition to calcium, the bones need constant physical stimulus and strain for growth. Gravity and muscle traction are absolutely necessary in order to keep the bones strong. Exercise provides these stimuli. Quiet children staying indoors might have weak bones. Astronauts on space flights do "resistant exercises" to prevent their bones from losing calcium and becoming weak because of the lack of gravity for even a short period.

Lack of sunshine exposure could also be dangerous. About one-half of the vitamin D circulating in our body enters as raw material, previtamin D, which has to be converted to vitamin D in the skin by the action of ultraviolet rays in the sunshine. Children always staying indoors may therefore have only one-half the amount of vitamin D compared to their friends who play under the sun. Vitamin D is very important for absorbing calcium from the gut, so that vitamin D deficiency invariably leads to calcium deficiency.

It would be fun to be in a spaceship, floating everywhere without the resistance of gravity pulling us down as it does on the earth. Without gravity, the world record for the high jump would be easily broken. To adjust to this environment without gravity, however, is no easy job. The first thing the early astronauts noticed when they returned to the earth was weakness of their muscles and then weakness of bones. Gravity, which prevents us from flying like birds, apparently does us some good. This was evident for the first time when men experienced a state without gravity. After one month in the spaceship, behaving like a superman, an astronaut could not even walk on earth because of his weak muscles and bones. When we stop using the muscles and bones, it is surprising how rapidly they become weak and thin. This is called *disuse atrophy*.

In order to prevent such a disaster, the new astronauts constantly exercise themselves in the spaceship. They have to create resistance against their muscles with various devices which substitute for gravity. If they are successful in simulating natural gravity in their daily resistance exercise, giving the bones exactly the same strain and load they are used to receiving on earth, the bones are kept as strong as ever and no trouble is expected.

About 30 years ago, polio was a terrible disease all over the world, including the United States. When a young man or woman was infected and extensively paralyzed, the bones were immobilized, just like the crew in the spaceship. The bone rapidly lost calcium and became very thin and weak. Large amounts of calcium and phosphorus released from bone passed through the kidney, producing kidney stones, thereby damaging the kidney function. This is another example of the importance of exercise for maintaining bone strength.

People who do vigorous exercises, jogging, tennis, swimming, and the like, are fortunate enough to be able to build up strong bones. They need a lot of calcium to add to the bone. In addition, they lose some calcium in the sweat, so that abundant calcium intake is necessary for sportsmen and sportswomen. During the growth period, adequate sports and calcium intake are especially necessary to build a strong body, because these two, hand in hand, help our bones to become strong.

Muscles do not contract in the absence of calcium. The role of calcium in muscle contraction is so important that it is impossible to discuss the problem of muscle contraction without referring to calcium. For powerful and timely muscle contraction, the calcium message should be good. No wonder athletes all over the world have started to take calcium to support their training and to improve their record. The calcium signal—indispensable for muscle contraction and nerve function—can be maintained only with adequate calcium intake, strong bone and vast inside/outside difference of muscle and nerve cells.

Bones and teeth store 99 percent of the calcium in our body. Evidently, sufficient calcium is a prerequisite for strong teeth. The mother gives so much calcium away to the baby that she frequently loses calcium from the bones and teeth after birth, unless enough calcium is supplemented. With advancing age, calcium deficiency becomes more pronounced because of the decreased efficiency of intestinal calcium absorption. This is one of the reasons elderly people lose their teeth. While multiple factors are responsible for the loss of teeth in aging, including *periodontal diseases* or changes in the tissue surrounding the teeth such as *alveolar bone disease* and *alveolar pyorrhea* due to bacterial infection, sufficient calcium intake contributes to strong alveolar bone and *gingiva* (gums), and increases resistance to infection.

Although every attempt is made to keep serum calcium level constant, minor fluctuations do occur. When calcium-rich food is ingested, serum calcium rises a little immediately following absorption. With even a slight increase of serum calcium parathyroid hormone secretion is inhibited. If the serum calcium level decreases, even slightly, parathyroid hormone secretion is stimulated. Therefore, consistently eating a low-calcium diet would result in a constant stimulation of parathyroid hormone secretion. A calcium-rich diet, on the other hand, would provide a "break" from such stimulation.

After months and years of constant stimulation of the parathyroid glands due to low calcium intake, the glands become enlarged and secrete parathyroid hormone without much stimulation. This is called *secondary hyperparathyroidism.* There is always abundant parathyroid hormone in the blood of subjects with secondary hyperparathyroidism. The bone is stimulated by parathyroid hormone, the number of bone-resorbing cells (*osteoclasts*) increases, bone is resorbed, and calcium is released into the blood stream. Thus when alveolar bone is gradually resorbed, the teeth are unable to stay in their place, and are lost one by one. Also, in a disease called primary hyperparathyroidism, when parathyroid hormone is spontaneously secreted in excess, many teeth are lost.

Among the natives of New Guinea, elderly people become unable to eat as they lose their teeth, because they do not know how to make the food soft or how to fix poor teeth. When they lose their teeth, they have to starve to death. Keeping or losing the teeth is thus a problem of life and death. Adequate calcium intake is also important for keeping the teeth in place.

As to the adequate amount of calcium in food, opinions of nutritionists vary. The calcium requirement is usually estimated based on balance studies, comparing total calcium intake or calcium in food with total calcium output, that is, all calcium excreted in urine and feces. This is like home economics. A family has to spend a certain amount of money for food, clothes and housing. In case the total income is less than the total of these expenditures, the family's economy is in a negative financial balance and cannot get along. When the income is larger than the expenditure, the economy is in a positive balance, making it possible to save some money. In order to determine the calcium requirement, nutritionists measure the amount of calcium in the food and then calcium output in urine and feces using a metabolic balance technique. The amount of calcium necessary and sufficient to produce a positive calcium balance is the calcium requirement. Our body is able to adjust to various levels of calcium intake, excreting less when a smaller amount is taken, like a thrifty homemaker managing the home economy with whatever is available. However, the fact that the body maintains such a calcium balance does not necessarily indicate that the body is getting an ideal calcium intake. If more calcium were available, the body could have used it for many good things like strengthening the bones, while still keeping the calcium economy in balance. A balance may thus be achieved at various levels. As most of us are quite sure that we could spend more money effectively for a good purpose, the body can always spend more calcium. We should not restrict calcium intake because the body is already in balance, because the body wants more and can use it.

The amount of nutrients in food is also related to calcium availability. Even if the same amount of calcium is contained in the food, the amount of calcium which is actually absorbed from the gut and used for the body may be different, depending on other components of food. When a constant amount of calcium is contained in the food, and more protein is taken as food, urinary calcium excretion is increased and less calcium is left in the body. Even if the same amount of 600 mg of calcium is taken, enough calcium might be retained in the body when protein intake is low, but the same amount of calcium may prove insufficient when a larger amount of protein is taken.

The relationship between calcium and phosphorus in food is also complex. Too much phosphorus in the food interferes with absorption of calcium from the gut, because calcium is readily bound to phosphorus, producing a precipitate which cannot be easily absorbed. The utilization of calcium might be decreased in the presence of an excess amount of phosphorus. Children nowadays take abundant protein and phosphorus. Even at the same levels of calcium intake, children taking large amounts of protein and phosphorus are thus more at risk than those taking smaller amounts of these. However, the effects of phosphorus and protein on calcium balance may depend on the type of food and the relative amounts of these two nutrients. In combination, they tend to somehow attenuate the effects of each other. However, studies in this area are often contradictory. This problem is especially important in the United States, where children are taking abundant protein and phosphorus, but maybe not as much calcium. It is quite possible that many American children are not getting enough calcium for their growth and health.

As mentioned earlier, a certain amount of exercise with solar exposure is necessary to make strong bones. In order to overcome the interference by protein and phosphorus, a sufficient amount of calcium should be taken as the most important prerequisite for adequate growth.

(2) Brain function and calcium

Many organs are found within our body, fulfilling their assigned functions. The stomach and intestine are responsible for digestion and absorption, the liver for metabolizing food and the kidney for excreting the metabolites from the body. These and many other organs are all necessary for the health of human body. Even if some of the functions appear simple, the action of each organ is quite complex, and cannot be simulated completely even by the most sophisticated apparatus currently available.

Among various organs of the human body, none is as complex as the nervous system, especially the brain. Human beings are distinguished from other living organisms by their exceptionally well developed brain function. By standing on two legs, human beings for the first time became able to support the heavy brain, and this contributed to a tremendous development. How does calcium relate to brain function?

When the calcium level in the blood falls, muscle cramps appear and remarkable changes occur in the nervous system, especially in the

Fig. 11 Nerve excitability rises when serum levels of calcium and magnesium decrease, and falls when these levels increase.

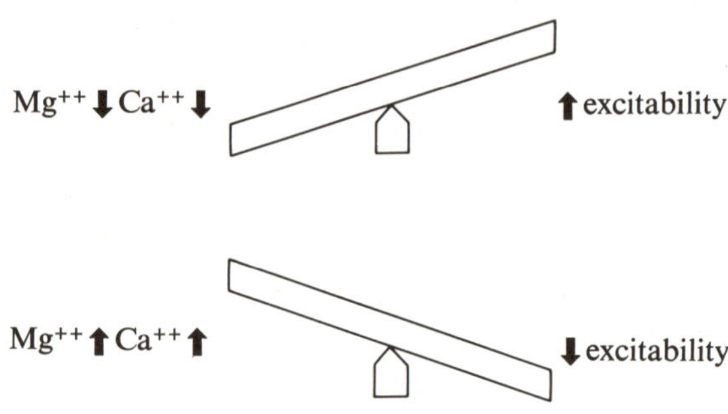

brain which is the central performer of this system. When the calcium level in blood becomes low, the nerve is more easily excited and becomes more unstable. Excitation in the physiological sense should not be confused with excitement of the mental function such as the one seen when one becomes angry. Whenever an electrical activity is generated in the nerve, the nerve is excited, and such excitation is transmitted, as an electrical current running through a wire. In the case of a telephone, a message is transmitted through a wire to a distant place. The message is transmitted by electromagnetic waves through the air for radio and TV.

When the dial of a radio is turned to increase the sensitivity, the voice of the announcer sounds louder, but it also picks up a lot of noise, making it more difficult to hear the voice of the announcer clearly. When the sensitivity is lowered, the extraneous noise disappears, but the voice of the announcer also becomes weaker. Calcium in blood is just like the volume-adjusting dial of a radio to control the sensitivity of the nerve and brain to various stimuli. When blood calcium is lower, the nerve is too easily excited, with more electric current running through in response to a small stimulus. Such a person is nervous and sensitive to small things. It may be difficult to get along with a person with a low blood calcium level in the blood.

On the contrary, when the serum calcium level is high, the nerve is not easily excited. The person with a higher than normal blood calcium level is dull, sleepy and not easily aroused. Normal blood calcium is therefore important for normal nerve function. Diets containing little calcium tend to cause a slight decrease in blood calcium. They might thus have a subtle effect of increasing brain and nerve excitability, although such a transient fall of serum calcium is immediately corrected by secreted parathyroid hormone, as long as the parathyroid glands are functioning normally. When parathyroid function is even a little lower than normal, a calcium-deficient diet may cause a decrease of calcium in blood, prompting nerve hyper-irritability.

The brain is the site of our heart and mind. The antenna of our mind is extended everywhere to catch the words of persons we talk to, events and experience. In case the calcium level in blood is low, various minute trifles of our daily life trouble us so much that we become restless and irritable.

Paranoia is an abnormal state of mind in which everything others do seems to be disturbing. Things others are saying innocently may sound to such a person as if they were speaking ill of him or her. When the blood calcium level is too low, people may occasionally experience such a condition. What is really nothing for a person with a normal state of mind might be quite disturbing for such a sensitive person. This is by no means a healthy state of mind, but an abnormally irritable state. Painful muscle cramps called *tetany* might also develop when the blood calcium level is too low.

Some of the people around us are quite sensitive at times. Only a word without any harm meant might prove quite traumatic to such a person, who may become suddenly angry. The mood of such a person is quite changeable, like the weather in the spring. Such an unstable, restless and irritable character may be seen in persons with a decreased blood calcium level. However, people may be irritable even with a normal blood calcium level. Nevertheless, when a person is irritable because of a low blood calcium, return of a normal blood calcium level would restore the individual to a healthy and normal frame of mind. The normal range of serum calcium is approximately between 9.6 and 10.2 mg/dl. Serum calcium levels between 8.6 and 7.6 may make people somewhat irritable, and serum calcium below 7.6 would almost always causes some emotional abnormality.

If you are at times jumpy or irritable, it might be a good idea to have your blood calcium measured. If a person could not get along with colleagues and lost his or her colleagues and lost the job because of a low blood calcium level, it would be most unfortunate. A healthy spirit dwells in a healthy body, with normal blood calcium.

Blood calcium level becomes low, not because the amount of calcium in food is insufficient. Even if not enough dietary calcium is taken, blood calcium is kept constant at the sacrifice of the bone as long as parathyroid hormone and active vitamin D are available. In addition, during the day-to-day control of serum calcium levels, there may be a slight and transient fall of serum calcium which escapes routine measurement. After signaling parathyroid hormone secretion, the serum calcium always returns to normal, as long as the parathyroid glands respond normally. Parathyroid hormone further signals the production of active vitamin D in the kidney. In case the parathyroid glands do not function normally, or if vitamin D is deficient, serum calcium continues to be decreased and cannot be restored to normal quickly. In case either of these is insufficient, blood calcium immediately lowers even if the dietary calcium supply is adequate. At such times, various abnormalities in the mind and mood can occur.

In order to restore the blood calcium to normal, vitamin D should be taken, since it can do a part of the work of parathyroid hormone. Taking calcium alone would not be sufficient in such cases.

Too high a serum calcium level, on the contrary, may be caused by secretion of too much parathyroid hormone or by taking too much vitamin D. In this case, a person may be depressed, not interested in his or her surroundings, and sleepy. Muscle strength may also decrease.

Epilepsy is a convulsive seizure of the whole body caused by generation of unusually high electric activity in the brain. *Grand mal* or major attack is a generalized convulsion with loss of consciousness; *petit mal* or minor attack is a limited cramp and psychomotor seizure consists of only psychological abnormality. Some of these are detected only by electroencephalography. Abnormal electricity is generated by various causes—birth defects, head injury and others.

Among children suffering from convulsions, many have a low blood calcium level because of insufficient secretion of parathyroid hormone. By giving them the active form of vitamin D to normalize blood calcium levels, the convulsions may decrease or even completely disappear.

Measurement of calcium in the blood is a very simple test requiring only about 0.5 cc blood. Those who are suffering from convulsions should measure the calcium level in blood at least once to see if there is any indication that treatment with the active form of vitamin D is advisable.

If properly used, blood tests can be very revealing. Ever since Hippocrates, physicians have attempted to observe every sign and symptom of the patient with extreme care. The association of jaundice with swelling of the liver is an example of a facial sign indicating a disease, in this case liver disease. A peculiar smell of the breath also helped to diagnose the disease before its active development.

The "battery" of biochemical tests on blood, physical signs such as the size, consistency, surface property and tenderness of the liver were the only information available to physicians for diagnosis, prognosis and decision for therapy. No wonder many years of experience and skill were necessary to be able to make accurate diagnoses. With the advances in modern biochemical techniques, it became possible to conduct many tests simultaneously with less than 1 ml of serum in an autoanalyzer. The combined results of all these tests have made it possible to accurately diagnose the condition, predict the outcome and help to decide upon the treatment. A physician's skill and experience are of course invaluable, but today, modern physicians are fortunate to be able to start where the physicians in the past stopped because of the limitation of the available techniques. A blood test is therefore quite important. Measurement of blood calcium is especially important. Low blood calcium in persons with epilepsy might also be the result of long-term drug administration in efforts to control the convulsions.

Fig. 12 Midwife's position in tetany due to hypocalcemia.

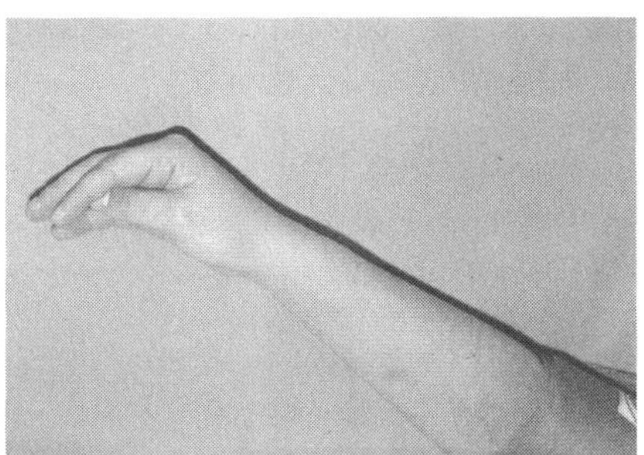

Decrease of calcium in the blood may cause muscle cramps or tetany, due to increased excitability of the muscle as well as the nerve. Muscle contraction is thus facilitated under slight stimulus. The midwife's hand is a characteristic position of fingers in tetany.

Children frequently suffer from cramps and no doubt their mothers are worried. Many of the emergency cases in pediatric service are related to cramps. Not all these children have low blood calcium, but it may be worthwhile to measure it, just in case.

Many of these cramps spontaneously disappear but some persist as habitual cramps or epilepsy, which may endanger normal mental development and even life. In case low blood calcium is responsible, treatment with active vitamin D is quite effective in normalizing blood calcium and getting rid of the cramps.

Cramps are caused by a sudden increase of electrical current passing through the nerve, causing muscle contraction. Decrease of blood calcium is one of the causes of such an unusual excitation of the nerves. Children are more susceptible to nerve excitation than adults because their nerves are more active and sensitive. A portion of calcium in blood is bound to protein and another portion remains free and ionized. When blood becomes slightly alkaline, the ionized portion of calcium decreases to sensitize the nerve excitation. Breathing too hard may drive carbon dioxide gas away from blood making the blood alkaline and causing cramps. High fever causes rapid breathing and therefore may also cause cramps. Newborn babies frequently show low blood calcium levels and cramps, and this alarms the parents. The parathyroid glands of some of these babies are not yet ready for the new life outside the mother's womb, and the glands do not secrete enough parathyroid hormone. After a few weeks, however, the parathyroid glands start to function normally, and in most cases blood calcium levels normalize. Meanwhile, calcium and active vitamin D supplements will raise blood calcium levels to carry them through without cramps.

3. Calcium and the Body's Self Defense

(1) Calcium and the immune function

Our body is constantly defending itself from enemies coming from outside, to achieve a quiet and fruitful life inside. Against the bacteria and foreign bodies invading us from outside, the immune system in our body acts as a guard or policeman. As is well known from Edward Jenner's vaccination against smallpox and the Salk vaccine against poliomyelitis, immunity is a method to escape epidemics, like a tax exemption privilege. Jenner inoculated his own child with cowpox, which caused a mild disease but prevented him from further infection by real smallpox. This was the beginning of the history of immunology.

When widespread viral infection attacks a community, everybody suffers from the infection, like paying tax, although the response of individuals might vary. Those who have gained immunity get along without any tax, like a man who is exempt from the tax. Immunity means you don't have to pay tax. What is the role of calcium in immunity?

When bacteria or virus invade our body from outside, *macrophages* will catch them at first and alarm the lymphocytes through a complex network. Lymphocytes transmit the information to plasma cells which produce antibodies to attack these invading organisms, to kill them or at least inactivate them to prevent further spread of the organisms throughout the body.

These cells responsible for immune function transmit their signals with the help of calcium. Macrophages are the cells freely moving in search of the invading organism. When they find their enemies, they extend the *pseudopod* or leg-like projection to eat them up. The movement of a macrophage is started by the entrance of calcium from outside across the cell membrane against the vast concentration gradient. A cell has its own bones and muscles which are called the cytoskeleton, consisting of microtubules and microfilaments. These structures maintain the shape of a cell just like the skeleton keeps the shape of our whole body. For the movement of a cell, calcium causes contraction of these "muscles in the cell," exactly like the contraction of the muscles of our arms and legs. When macrophages cannot completely eliminate the invaders, they send messages for

Fig. 13 Calcium and cell movement.
Calcium is required for each movement of the cell. Flagellum is a single filament which guides and moves the organism, through the action of a contractile protein regulated by calcium: A similar control by calcium is exerted in cilia. The pseudopod is a "false leg" of an amoeba or white blood cell and is also controlled by calcium. Change in cell shape is controlled by the cytoskeleton, "bones of a cell," and this cytoskeleton consists of Ca-controlled contractile proteins.

Calcium and Cell Movement

Ca (–) ⟶ Ca (+)

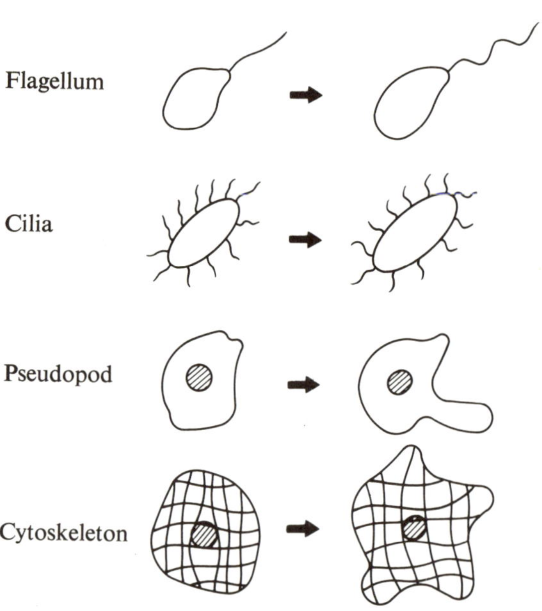

more help. Release of such messengers, one of which is called *interleukin 1*, requires calcium, as do other hormone secretions from endocrine cells. There are several kinds of lymphocytes for different duties, like the police system. Some of them, called *helpers*, encourage others to go ahead while the ones called *suppressors* are more cautious; Their job is to make sure others are not overzealous in defending the body. As police officers carry walky-talkies, calcium is the means of communication and the signal for these various kinds of lymphocytes and plasma cells. When calcium is deficient in the food, parathyroid hormone may be secreted. One of the functions of parathyroid hormone is facilitation of the entrance of calcium into the cell to cause a rise of intracellular calcium. However, the action of parathyroid hormone so far has been demonstrated in only certain types of cells. The concentration gradient across the cell membrane of the macrophages and lymphocytes might therefore be blunted. Consequently, these cells cannot hear each other or move well, nor can they efficiently eat up or attack the invaders. The signals for the network among these cells may become hopelessly confused.

In *autoimmune diseases*, the immune system of the body which is supposed to act against invaders actually attacks part of one's own body by mistake.

Several things are responsible for such confusion. Heredity strongly influences the mode of action of the immune system. A tendency to suffer from autoimmune disease may run in the family. Some strains of animals showing such tendencies are used to search for the causes and treatment of such diseases. Females are usually more frequently affected by autoimmune diseases than males. The mistaken immune response may also occur after some of the body cells, like thyroid cells or liver cells, have changed their shape and property. As long as certain body components are confined in a closed compartment, without direct contact to circulating blood, such as the thyroid follicle or the inside of the eyeball, they may remain non-immunogenic. Once the boundary for such a compartment is broken, as in an eyeball injury or in *thyroiditis*, antibodies against these substances are formed and autoimmune reactions proceed often involving the other eye or the rest of the thyroid gland. A policeman may shoot another policeman by mistake if this policeman appears or behaves exactly like a bank robber. When the cells go out from the compartments where they are usually confined, a mistake might also occur, like a police officer who goes to a place where he is not expected and gets shot by mistake.

An allergy is also a harmful event related to the immune mechanism of the body. Reaction between the *antigen*, the target for the immune defense system, and *antibody*, the weapon used by this system, may be exaggerated to cause a disease such as hives or asthma. Autoimmune diseases frequently involving connective tissue and blood vessels are called *collagen diseases*. *Systemic lupus erythematodes*, *polymyositis*, *polyarteritis nodosa*, and *scleroderma* are examples.

Among the lymphocytes in the immune system, suppressor cells are supposed to inhibit excessive and erroneous activity of other lymphocytes. Loss of this inhibitory activity of suppression lymphocytes may represent one of the causes of autoimmune deficiency phenomena. In case a calcium deficiency prompts the loss of function of suppressor lymphocytes, the mistaken attack by the overactive lymphocytes cannot be corrected. Autoimmune diseases, like a riot or civil war within our body, may thus be started. In many patients with osteoporosis, suppressor cells are also decreased, possibly because of calcium deficiency. This is corrected by active forms of vitamin D.

The immune mechanisms in our body are so complicated that the action of calcium alone cannot explain everything. Recently, the active form of vitamin D was found to have a controlling function on the immune system. Children who cannot make active vitamin D and patients with chronic renal failure unable to make the active form of vitamin D are known to suffer from repeated infections. In addition to being a key substance in calcium absorption from the gut, vitamin D can change the property of blood cells to give them the functions of macrophages. Vitamin D increases the amount of calcium binding protein in various tissues, probably influencing the intracellular free calcium concentration. It may also act on the cell membrane directly to change the *phospholipid* constituents of the cell membrane, thereby influencing the rate of calcium entry into the cell.

Sufficient intakes of calcium and vitamin D therefore help to prevent and treat diseases caused by abnormalities of immunity. Sunbathing in summer is said to prevent the common cold in winter. Ultraviolet rays in the sunshine increase vitamin D synthesis in the skin, and this might be good for immune function. No wonder the people living in northern countries are anxious to get as much sunshine as possible during the short summer months.

The immune function is at work in fighting various infections such as the common cold, bronchitis, pneumonia, *cholecystitis*, *enteritis*, and pyelitis caused by bacteria or virus. When sufficient immune function is not ready, even a potent antibiotic cannot save the body from infection, because antibiotics alone seldom kill the bacteria completely and the remaining bacteria have to be killed by the body's own immune mechanism. When one becomes older, the immune function gradually deteriorates. Pneumonia is usually easily cured in young people, but may prove fatal in elderly people for this reason.

As one becomes older, calcium absorption from the gut declines, and, furthermore, elderly people have a tendency to take less calcium and vitamin D containing food in general. Less exposure to sunshine may even accelerate vitamin D deficiency because less vitamin D is synthesized from previtamin D in the skin. It is of interest that the skin contains a small amount of 1,25$(OH)_2$ vitamin D and the content of this active vitamin D decreases with advancing age. Parathyroid hormone thus cannot help becoming active in elderly people, because of an augmenting vitamin D and calcium deficiency. Calcium goes into each cell including those participating in immunity by the action of parathyroid hormone. The consequent blunting of the inside/outside differences of calcium concentration makes the messenger system of the cell less efficient. Calcium deficiency may therefore lead to a decreased resistance to infection.

(2) Rheumatoid arthritis and calcium

Rheumatism simply means pains in the joints including the back, shoulder, fingers, and so on, as well as muscles, but the real causes are quite diverse and complicated. On the other hand, *rheumatoid arthritis* usually affecting young to middle-aged women, is a distinct disease with a typical chronic course, leading to severe joint deformities, due to disturbed autoimmunity. *Gout*, caused by uric acid deposition in the joint, may also belong to this group of diseases called rheumatism, though its cause is entirely different from that of rheumatoid arthritis.

Rheumatic fever, mainly seen in children, is another disease with multiple joint involvement, a consequence of *hemolytic streptococcal infection.* Since so-called collagen diseases are related to a confused immune control system, calcium deficiency might be involved in the etiology, by interfering with the function of the lymphocytes and their messenger system. Susceptibility to these diseases is somehow influenced by heredity, but environmental factors are also important. Females are more frequently affected than males by these diseases, except in the case of polyarteritis. Such sex differences in the prevalence of autoimmune and collagen diseases cannot be fully explained at present. Males and females have different chromosomes, providing different hereditary backgrounds. Estrogens providing female characteristics may also have some influence. On the other hand, females may be in more danger of calcium deficiency because they tend to eat less calcium-containing food and have to spend a large amount of calcium during pregnancy and lactation. This must be one of the reasons why females suffer from bone loss and osteoporosis more frequently than males.

In a similar way, females might be expected to suffer more frequently than males from dysfunction of immune cells, including suppressor lymphocytes. This would occur as the calcium deficiency blunts the inside/outside gradient of calcium across the cell membrane, and could encourage the development of autoimmune collagen-vascular diseases.

Rheumatoid arthritis, frequently seen in adult females, is a dreadful incapacitating disease. This disease is characterized by morning stiffness and pain of many joints all over the body. Eventually many joints, especially the fingers and wrists, may become deformed. In this disease, the bones around the joints become especially thin and weak. Pain and swelling of the joint restricts the movement of the bone and joint, and this might contribute to such thinning, but the degree of thinning is much greater in rheumatoid arthritis than in *osteoarthritis*. This might be related to calcium deficiency which frequently occurs in patients with rheumatoid arthritis. In a joint cavity that is affected by rheumatoid arthritis, many kinds of cells are active in producing materials modifying the immune function, and these might have an effect on the nearby bone.

In patients with rheumatoid arthritis, the action of vitamin D is decreased and calcium absorption from the gut is low. With constantly aching joints, a patient's appetite is usually poor so that nutritional calcium and vitamin D intake tend to be limited. Pain in the joints, moreover, prevents these patients from going out in the sun very often. This would again deprive them of the precious opportunity of receiving ultraviolet rays on the skin for vitamin D synthesis. Because of the insufficient action of vitamin D, the calcium level in blood is slightly decreased in many patients with rheumatoid arthritis, and parathyroid hormone is apparently produced in excess. The bone is probably losing calcium under such conditions.

A vicious cycle is probably taking place in a similar way in many other diseases. Calcium deficiency causes an immune abnormality. Rheumatoid arthritis develops because of an autoimmune phenomenon based on some derangement of the immune mechanism. As the disease advances, calcium and vitamin D deficiencies proceed, further aggravating the condition.

In order to interrupt such a vicious cycle, sufficient amounts of calcium should be taken and active vitamin D might be supplemented. This treatment has been reported to improve the bone of patients with rheumatoid arthritis. In some patients, pain and swelling of the joints improved, making daily activities easier.

(3) Kidney disease and calcium

Many of you may have experienced sore throats, with swelling of the tonsils. *Tonsillitis* sometimes leads to *nephritis* and protein leak in the urine. When the immune system of our body reacts to hemolytic streptococcus, a common pathogenic organism for tonsillitis, it might also attack our own kidney. This is because the basement membrane of the *glomerulus*, where the blood is filtered to produce urine, is easily mistaken for the bacteria by the immune system. If calcium deficiency is somehow related to immune dysfunction and the development of autoimmune disease, nephritis would also be an example.

In the glomerulus, a tangled ball of capillary blood vessel in the kidney where the blood is filtered to produce urine, a tissue called *mesangium* is found which assembles the capillary blood vessels. When calcium is deficient, calcium comes out of the bone and may enter the mesangial cells. Mesangial cells are like smooth muscle cells on the vascular wall, capable of contraction when calcium enters the cells. The consequent contraction of these mesangial cells accelerates the advance of *glomerulonephritis* from an acute disease to a chronic one. Continuation of such changes over the course of many years may lead to renal insufficiency and renal failure. Calcium deficiency may thus play a role in the development and advance of renal disease. Adequate calcium and vitamin D intake are necessary to help prevent the progress of renal disease.

As the kidney disease progresses and much less intact kidney tissue remains, chronic renal failure occurs. Before the development of *hemodialysis*, the filtering of blood to remove various metabolites not required for the body, patients with progressive renal failure could do nothing but await death. Hemodialysis has saved the lives of patients with chronic renal failure.

It has now become possible even for patients with no kidneys at all to live with regular hemodialysis, because hemodialysis performs the most important function of the kidney, blood filtration.

The kidney, however, has another important function which should not be forgotten. It is the only place where the active form of vitamin D or $1,25(OH)_2$ vitamin D can be produced. Evidently the dialyzer cannot take the place of the kidney in synthesizing active vitamin D.

Therefore, the most dreadful complication of chronic renal failure is vitamin D deficiency. In patients with chronic renal failure, intestinal calcium absorption is markedly decreased. Blood calcium concentration falls and more and more parathyroid hormone is secreted.

Continuously stimulated, the parathyroid glands increase in size, sometimes dramatically, even thousands of times their original size. Although at normal levels of secretion, parathyroid hormone alone cannot act very well on the bone to withdraw calcium, at tremendously high levels of parathyroid hormone, even in the absence of active vitamin D, calcium is gradually lost from the bone. The decrease of active vitamin D also inhibits bone calcification, which often leads to the development of *osteomalacia*. *Renal osteodystrophy* is a complex bone disease seen in patients with longstanding renal failure, consisting of *osteitis fibrosa* and osteomalacia. Aluminum is frequently deposited in the bone, interfering with calcification and aggravating the osteomalacia.

Calcium coming out of the bone inevitably enters soft tissues such as blood vessels and nerves. In patients under long-term dialysis with inadequate vitamin D supplementation, the function of the nervous system is decreased and the speed of transmission of stimuli through the nerve is decreased (D. A. Goldstein et al 1978). Muscle weakness and decreased sensation may be the consequences. The function of the brain may also fail as in dementia. Calcium and aluminum are increased in the brain, spinal cord and peripheral nerves of patients with chronic renal failure (A. I. Arieff et al 1974). In patients with dementia and nervous disease associated with renal failure, the first step in treatment should be to administer active vitamin D to stimulate gut calcium absorption.

(4) Bronchial asthma and calcium

Bronchial asthma is caused by contraction of bronchial smooth muscles, leading to a narrowing of the airway. Allergies to pollen or other chemical substances usually trigger the chain of events leading to entry of calcium into the bronchial smooth muscle cells. Calcium antagonists tend to ameliorate the asthmatic attack although the exact mechanisms have not yet been fully elucidated. Bronchial muscle contraction may be inhibited by the inability of calcium to enter smooth muscle cells. Adequate calcium intake is recommended in asthmatic patients, because this would help to maintain an optimum inside/outside calcium balance across the smooth muscle cell membrane.

Calcium antagonists, a group of drugs which inhibit calcium entry into the cell, have been found to alleviate asthmatic attacks. When *corticosteroid drugs* are used to control the asthmatic attack, calcium is excreted in increasing amounts in urine. In the case of children, this may retard growth and weaken the bones. Therefore, calcium supplements along with vitamin D may be of help in these conditions. It should be clearly understood that calcium is the best calcium antagonist to block the entrance of calcium into cells. This is the Calcium Paradox. Sufficient calcium intake inhibits parathyroid hormone secretion and inhibits calcium entry into cells. Deleterious effects of corticosteroid in postmenopausal women are quite obvious. Calcium is lost from bone much faster in response to corticosteroid.

4. Calcium and Adults

(1) Osteoporosis and calcium

a. Osteoporosis, a "boneless" disease: Osteoporosis, a disease characterized by increasing numbers of "pores" or holes in the bone, is nothing but bone loss, or a "boneless disease." Everyone loses bone with age, because all of us are deficient in calcium. When such bone loss becomes so severe that the bones are no longer able to bear body weight or muscle strain, fracture is imminent. This is osteoporosis. When this age-bound decrease of bone mass is only mild, little difficulty arises, it is considered rather *physiological bone loss of aging*, and is distinguished from osteoporosis only by the degree of severity.

The sphinx of Egypt used to ask a question of passing travellers. "What is the name of the creature which walks on four legs in the morning, two legs at noon and three legs in the evening?" Whoever was unable to answer this question correctly, "It's us," was eaten up by the Sphinx. Babies crawl on all fours, adults walk on both legs, but elderly people require a cane because they develop osteoporosis, lumbago and roundback. This has been occurring since the ancient times of the mysterious Sphinx. In the traditional stories and fables, probably reflecting our lives in ancient times, old women are always using canes, while many old men are walking with straight backs. Women have probably suffered from osteoporosis more frequently than men almost since the beginning of mankind.

What is osteoporosis? It is defined as a decrease of total bone mass with preservation of the normal proportion of bone constituents to a degree which makes it difficult to support the body weight. Normal bone constituents include bone cells, minerals mainly consisting of calcium and phosphorus, and an organic matrix consisting of collagen and *glycosaminoglycan.*

Fig. 14 Normal and osteoporotic spine. Internal structure has almost disappeared in osteoporosis, producing a fragile bone.

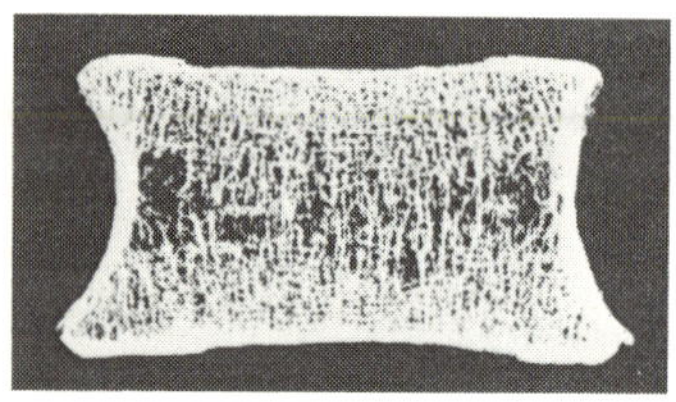

Normal Spine

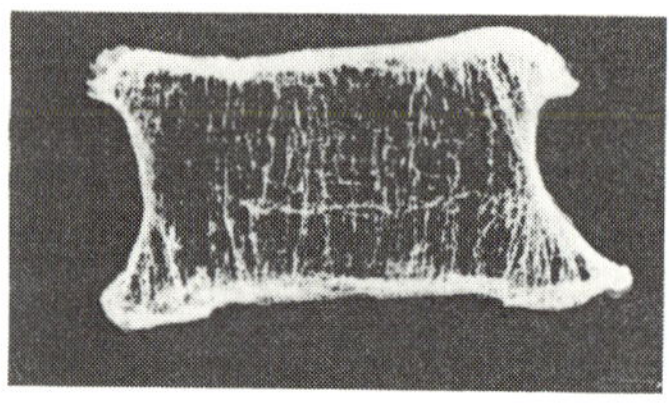

Osteoporotic

When the bone mass is decreased and the bones become too weak to support the body weight, fractures start to occur and the spine is compressed and deformed. This gives rise to round back, lumbago and backache, the main manifestations of osteoporosis.

Fig. 15 Most common fractures seen in patients with osteoporosis. From left to right, the most disabling hip fracture seen in the oldest age group, spinal fracture common in moderately aged people, and forearm fracture seen in younger age groups.

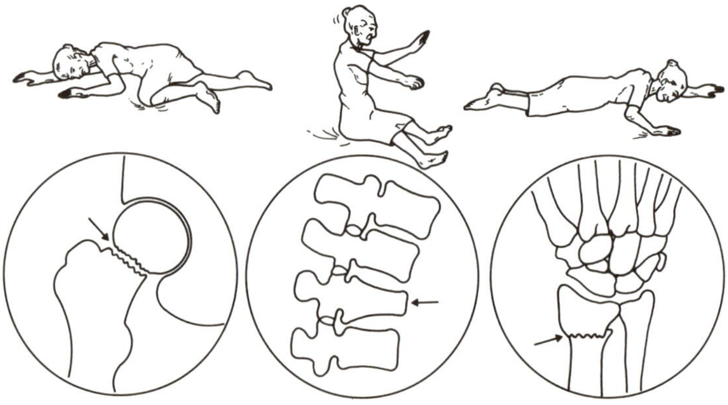

The three main sites of fracture in osteoporotic patients are: 1) forearm, 2) spine, and 3) hip. Forearm fractures occur at the earliest age, followed by spinal compression fractures which are the most common type of fracture in osteoporotic patients. Finally, fractures of the hip or neck of the femur usually occur in the oldest age groups and are the most dreadful, because they could completely incapacitate the patients, by keeping them bedridden for more than six months, if not treated promptly and adequately.

As a broader term, decrease of bone mass due to any cause may be called *osteopenia*. Osteopenia includes osteoporosis, as well as bone loss due to many other causes. Osteomalacia is distinguished from osteoporosis by a defect in the calcification mechanism, resulting in an increased proportion of uncalcified matrix or *osteoid*. Metastasis of malignant cells to the bone also differs from osteoporosis by the presence of an increased number of abnormal tumor cells.

Fig. 16 Prevalence of soteoporosis in Japan.
Osteoporosis increases with age, especially in females, everywhere in the world.

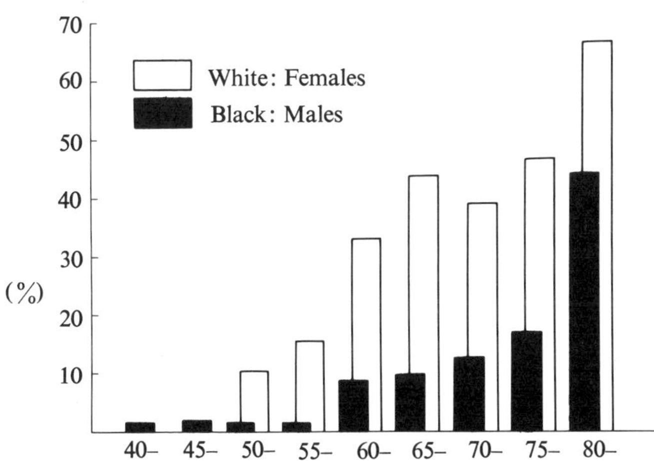

b. Who is at risk for osteoporosis? Females are always more frequently affected by osteoporosis than males. Although osteoporosis may affect anybody at any age, it is most frequently seen in women after menopause. Postmenopausal or senile osteoporosis may be the most prevalent disease of mankind.

Menopause occurs in women around the age of fifty. The menstrual periods stop and secretions of female hormones, estrogen and progestin, suddenly and drastically decrease.

Estrogen from the ovarian follicle not only properly maintains the womb and breasts in preparation for childbearing, but also controls the metabolism of blood vessels and bones, cooperating with other hormones. When estrogen suddenly drops out from the teamwork of hormones in the human body, a confusion appears causing manifestations of the so-called "change of life." Hot flushes are most characteristic, followed by other climacteric symptoms such as sweating, pins-and-needles sensation, stiff shoulders and difficulty in falling asleep. Autonomic nerve imbalance also occurs. These symptoms are frequently confused with neurosis. In women with climacteric symptoms, estrogen administration dramatically improves the symptoms. This is the best method to tell the difference between true climacteric syndrome and neurosis. Males also suffer from change of life or climacteric, but the change is not as abrupt as in women, because of a slower decline of male hormone secretion.

More serious consequences of menopause are loss of bone or osteoporosis. Estrogen deficiency can have deleterious effects on the bone within 10 years. This occurs regardless of the cause of estrogen deficiency, a natural menopause or the surgical removal of the ovaries due to a disease. Females develop osteoporosis several times more frequently than males.

Why does osteoporosis occur when estrogen secretion is insufficient? Estrogen appears to be necessary to store calcium within the body. Like a thrifty homemaker, estrogen tries to get along with as small a household budget as possible. He or she is always trying to minimize the withdrawal of money from the bank, in the same way as estrogen opposes the parathyroid hormone's action of withdrawing calcium from the bone. Estrogen also helps the thyroid to make more calcitonin, which acts on the bone to prevent calcium coming out of the bone by inhibiting the *osteoclasts*, giant cells which can break down and resorb bone. Estrogen also helps the kidney to make the active form of vitamin D which is most important in maintaining gut calcium absorption. Thus estrogen is important for saving calcium in the bone, thereby keeping the bone strong.

Why does estrogen want to keep the bone strong? Because the bone must be strong for safe childbirth. One of the most important missions of estrogen is to make sure a baby is safely carried through the pregnancy and delivery. The head of the baby must pass through the pelvic canal. If the bone becomes weak, the pelvis may be deformed so that the baby cannot pass through the pelvic canal to be delivered. It is no wonder, then, that once this important hormone, estrogen, decreases during the change of life, calcium is lost and the bone becomes weak.

Women usually eat less than men, because of lower energy requirements and smaller body size, on average. Young women especially eat less because of the fear of overweight. Thus women can potentially be more easily deficient in calcium and vitamin D than men. Vitamin D deficiency makes even less calcium available. This might be another reason why females suffer from osteoporosis more frequently than males. During pregnancy and lactation, females have to spend more calcium. Thicker and bigger bones would last longer before developing osteoporosis. Males usually have larger frames and therefore thicker bones than females, and this appears to be one of the reasons why males develop osteoporosis less frequently than females. It is well known that black women almost never develop osteoporosis. This may be because they have thicker bones in their youth and hence are more resistant to calcium deficiency. Blood calcitonin was reported to be higher in black women than in white women. This might also help to explain why black women are protected from osteoporosis. The amount of calcitonin in blood is usually higher in men than in women, and this may be yet another reason why osteoporosis occurs more frequently in women.

Fig. 17 Vitamin D metabolites in the blood of osteoporotic patients and normal subjects.
In osteoporotic patients, 25(OH) vitamin D was almost normal, but 24,25(OH) vitamin D was reduced and 1,25(OH) vitamin markedly decreased.

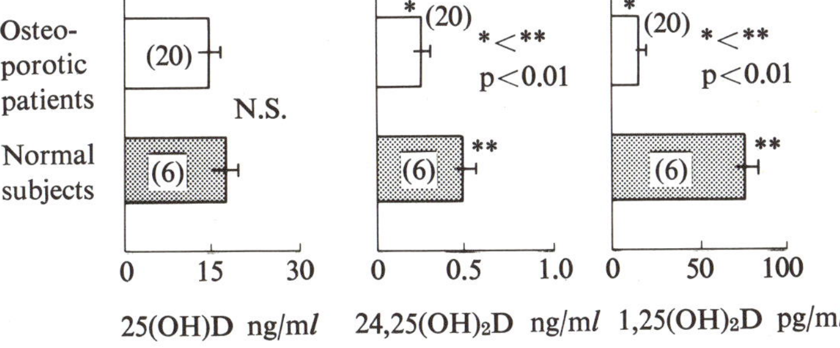

About one-half of the vitamin D used in our body is made from previtamin D in the skin with the help of ultraviolet rays in the sunshine. The northwestern side of Japan facing Siberia across the Japan Sea is buried under deep snow throughout winter, in sharp contrast to the southeastern side facing the Pacific Ocean which enjoys abundant sunshine all year round. The high central mountains dividing Japan from north to south make all the difference, and these two districts represent a rare example of sharp differences in climate between two nearby regions. A survey of osteoporosis among diabetics in Japan revealed a more frequent occurrence of osteoporosis in the northwestern than in the southwestern side. The frequency of osteoporosis was inversely proportional to the duration of sunshine in a month. Vitamin D made in the skin with the help of sunshine thus appears to play an important role in preventing calcium loss and osteoporosis. In addition to maintaining calcium balance, control of immune function by vitamin D might also be important in preventing bone loss and osteoporosis through its controlling action on bone cells. Diabetes itself unfavorably influences the production and activity of active vitamin D, because insulin is required for the synthesis of active vitamin D (Imura et al., 1986).

Age itself is a factor because everybody loses bone with advancing age. Physical activity is another factor, since loss of stress and strain to the bone is always associated with bone loss, called *immobilization osteoporosis.* Nutritional factors, especially calcium intake, are the keys for preventing osteoporosis. Vitamin D deficiency or intestinal diseases that cause malabsorption accelerate calcium deficiency, predisposing one to osteoporosis.

In people suffering from general malnutrition of protein and calorie, maintenance of normal bone mass cannot be expected. Vitamin C deficiency can cause a decrease of bone formation. In addition, since thyroid hormone also interferes with intestinal calcium absorption and increases bone resorption, secretion of excess thyroid hormone in hyperthyroidism is also associated with osteoporosis. Excessive smoking and drinking alcoholic beverages may interfere with intestinal calcium absorption. Excess adrenocorticosteroid hormones interfere with the action of vitamin D, and thereby decrease intestinal calcium absorption. They also tend to increase calcium excretion by the kidney and so further contribute to a negative calcium balance. In *Cushing's syndrome*, a disease with excess secretion of glucocorticoid from the adrenal cortex, osteoporosis is frequently found to occur even in the young. Corticosteroid treatment for many years can also produce growth retardation in children and osteoporosis in people of all ages.

Fig. 18 Prevalence of osteopenia among diabetics in various parts of Japan.
Osteopenia is more frequent in the northern part than in the southern part, and on the Japan Sea side than on the Pacific side.

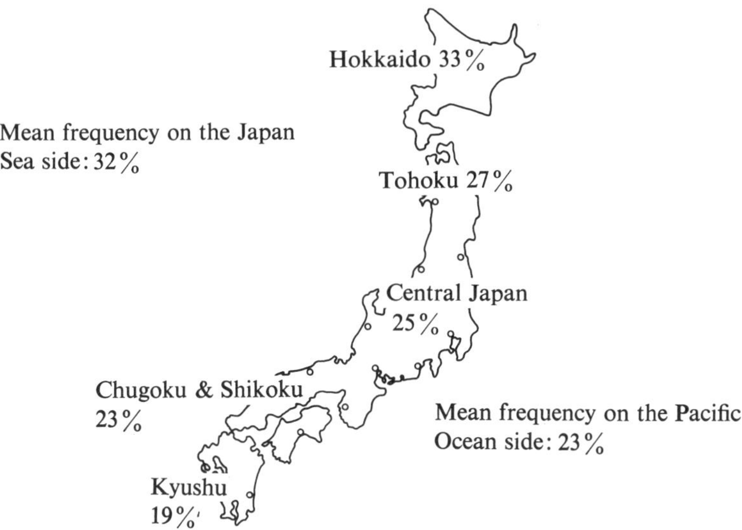

Osteoporosis from known causes such as from Cushing's syndrome is called *secondary osteoporosis.* When the cause is treated, secondary osteoporosis may be cured. The majority of osteoporosis, however, is primary osteoporosis, where the cause is unknown. This is also referred to as postmenopausal or senile osteoporosis and predominantly seen in women above middle age.

Osteoclasts are located on the bone marrow surface and *Haversian canals* are within the bone. *Osteoblasts*, on the other hand, are bone forming cells, located on the *periosteal* or outside surface as a thin layer. Osteoblasts first make a collagen framework and network of the bone and then help to calcify it, making a complete, strong bone.

What is bone doing in our body? Along with the teeth, bone makes up the hard tissue of our body. Can we imagine what would become of our bodies without bone? We would become like jellyfish. The

Fig. 19 Very active osteoclastic bone resorption. The giant cells with many nuclei are osteoclasts.
Slide taken from a patient with parathyroid carcinoma secreting a tremendous amount of parathyroid hormone.

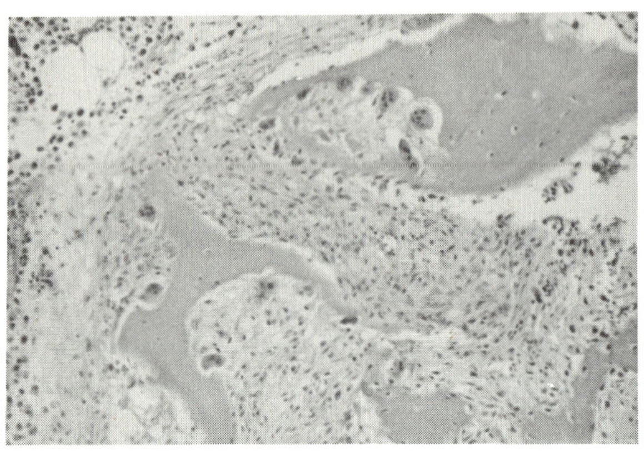

first and most evident function of the bone is to support and protect our body. If the skull were not strong enough, the mortality rate from traffic accidents would soar. We could not even move ourselves around properly without the support of the bone, because the two ends of all muscles are attached to the bone by tendons. By contraction of a muscle two bones may be pulled together or form angles. Every movement we perform thus depends on bones and the muscles attached to them.

The second function of the bone is to serve as a huge calcium reservoir, storing 99 percent of the calcium in the human body in a readily available form. Thanks to this enormous reservoir, serum calcium can be maintained at a precisely controlled level. Bone is constantly being made (*bone formation*) and also constantly being destroyed (*bone resorption*). Osteoclasts are giant cells with many nuclei and are mainly responsible for bone resorption.

When bone is resorbed by osteoclasts, all the organic and inorganic consitutents of the bone are released from the bone into the blood. Among these constituents, calcium is the largest component. Bone structure/content is thus based on the balance between bone formation and resorption. As long as resorption predominates over formation, bone mass will gradually be decreased, because more bone is being removed than built up. On the contrary, when more bone is formed than the amount which is removed, bone mass will gradually increase, as it does during the growth period.

We eat calcium-rich foods like cheese and tofu, and drink calcium-rich products like milk every day. The more calcium we include in our diet, the more we increase our "margin of safety" for the bones. Every day we are also losing calcium through the feces and urine. Calcium in the feces includes that which escaped the intestinal absorption, as well as that which was excreted in the bile and secreted by intestinal cells. Sweat also contains some calcium, but much less than the amount of urinary or fecal calcium. If the total amount of excreted calcium exceeds the amount injested, our bodies would be losing calcium from the body. For example, if the amount of calcium excreted was 700 mg, but only 650 mg of calcium was available from food, then 50 mg of calcium would be lost from the body on a daily basis.

After a week, this would total 350 mg and 18 grams in a year. If we kept losing this amount of calcium every year, 360 grams of calcium would be lost from the bone in 20 years. Even if the bone is a huge calcium reservoir containing as much as 1,000 grams, this would be a substantial loss, leading to osteoporosis (Avioli, 1983).

As we become older, the gut absorbs less calcium, partially because the gut itself becomes older, with less efficient circulation of blood and lymph. Aging also results in decreased formation of active vitamin D, 1,25(OH) vitamin D, in the kidney. In addition, as one ages, less 1,25$(OH)_2$ vitamin D is found in the skin. Estrogen deficiency in women also diminishes the amount of 1,25$(OH)_2$ vitamin D synthesized. In addition, elderly people generally eat less and are often exposed to sunshine less regularly. All these factors contribute to a decreased intestinal calcium absorption. In order to absorb the same amount of calcium, therefore, elderly people must eat a larger amount of calcium-containing food than younger people. Estrogen administration improves calcium absorption, probably through an increased formation of active vitamin D. Another important cause of senile osteoporosis may be the decrease of the bone's ability to store calcium. Compounding this problem is a more frequent state of negative balance, partly due to decreased active vitamin D levels, leading to a fall in serum calcium, and concomitant rise in parathyroid secretion. More calcium would thus be withdrawn from bone, as the parathyroid hormone causes an increased activity for osteoclasts.

At present, the only known natural inhibitor of osteoclasts is calcitonin, a hormone produced by the body itself. However, this level of secretion also decreases in older age, especially in women, a condition leading to further activation of osteoclasts. When serum calcium rises, the secretion of calcitonin is stimulated, while parathyroid hormone secretion is inhibited. Calcium deficiency, on the contrary, decreases calcitonin secretion and increases parathyroid hormone secretion. Calcium deficiency thus appears to have a dual deleterious role on bone metabolism, depleting calcium stores and inhibiting calcitonin secretion, thereby indirectly stimulating bone resorption.

Thus the complex network of estrogens, parathyroid hormone, active vitamin D and calcitonin loses its control in older age especially in women, causing loss of calcium from the bone. This is osteoporosis from the viewpoint of calcium metabolism. Beneficial effects of calcium supplementation and administration of estrogen, active vitamin D and calcitonin may confirm the role of these hormones in the development of osteoporosis.

Besides functions of support and calcium storage, the third function of the bone is to maintain the bone marrow cavity. No one knows why the blood-making bone marrow is located in the middle of bone, in the bone marrow cavity. However there may be good reason for this location. The inside of the bone may be where calcium is most abundantly and readily available. Osteoclasts, cells mainly responsible for bone resorption, are also derived from the bone marrow. As we age, the bone marrow cavity becomes wider and wider, while the outside measurement of the bone remains unchanged. The result is a thinner bone with less resistance to the outside force.

c. How do we recognize osteoporosis? When the bone loss reaches a level which cannot adequately support the body weight, fracture occurs. Spinal compression and deformity produce backpain and lumbago. Human beings, walking on both legs and carrying an exceptionally heavy brain, are unique in their susceptibility to spinal deformities due to osteoporosis. If the hip bone is broken, and adequate treatment is delayed, the patient might become bedridden. Osteoporosis is one of the major causes of disability in the United States and the rest of the world. This tendency will no doubt become more pronounced as people live longer. The proportion of the elderly people in the whole population has been steadily increasing.

Lumbago and backache are the hallmarks of osteoporosis. In osteoporosis patients, lumbago and backache occur when walking, carrying something or standing for a long time. Resting usually somewhat alleviates the back pains. When a spinal compression fracture occurs due to a fall or to carrying excessively heavy loads, severe backache may suddenly appear and may last a few weeks. This is due to spinal compression. In time, such pain goes away as the crushed spine adjusts itself to the new position. After some time, another spine may undergo a crush fracture.

Children become taller as they grow, but elderly people, especially women, become shorter. As osteoporosis progresses, each of the 24 vertebrae—7 *cervical*, 12 *thoracic* and 5 *lumbar*—may lose height. Even if one vertebral body loses only 1 mm, the loss of height would amount to 2.4 cm. After menopause, women may lose as much as 0.3 cm in height every year. In 20 years, she may have become 6 cm shorter! A roundback, or bending of the back, may also occur. For some unknown reason, the anterior edge of the spine is weaker than the posterior edge, and is thus more susceptible to deformity and compression. Forward bending of the spine is thus produced.

Fractures frequently occur in patients with osteoporosis Fracture of the forearm near the wrist might occur when a person falls forward and protects herself by supporting the impact with both hands. This fracture is called *Colles' fracture*. Perhaps the most serious fracture is the hip fracture, which also occurs upon falling, usually in much older persons with more advanced osteoporosis. Unless promptly treated surgically by insertion of a wire or nail into the thigh bone, a hip fracture may keep a person bedridden for as long as six months. This is not only quite depressing to the patient, but is also a heavy emotional and financial burden to the family and society. Profound anemia usually occurs when a long bone is fractured because bone also acts as a blood vessel to carry blood and bone marrow. Senile dementia is frequently prompted by fracture of the hip and the subsequent immobilization.

It is therefore important to recognize the risk of osteoporosis before reaching such a serious stage whereby fracture of the hip is sustained. It is naturally a rule of thumb for any disease that the earlier the recognition of the disease the better off we are in the prevention and treatment. We should try to detect osteoporosis as early as possible.

When Dr. Wilhelm C. Roentgen discovered the X-ray about 100 years ago, the first X-ray picture he took was that of his own hand. By recognizing the bones of his fingers dimly outlined in his picture, he realized his accomplishment in enabling one to visualize the bones for the first time. Since then, millions of X-ray pictures of the bones have been taken. Indeed the X-ray has become such a useful diagnostic tool that physicians and surgeons may sometimes expect too much from it.

Bone mass measurement is one of those tasks difficult for the X-ray. The X-ray is a mixture of gamma rays with various wavelengths, just as sunlight is a mixture of light rays with various wavelengths. It is extremely difficult therefore to estimate how much X-ray is absorbed by bone or other tissue. In order to accurately measure the absorption of light, or *optic absorbency*, a *spectrophotometer* was produced. A mixture of light is broken down into its separate components, similar to the rainbow produced by passing light through a prism. When the monochromatic light with a definite wavelength is passed through a solution, it gets absorbed in proportion to the amount of the substance contained in the solution, and by measuring the intensity of the unabsorbed light, we are able to tell exactly how much of the substance has been retained in the solution.

In the past, X-ray pictures were used to measure the amount of bone present, but it was soon learned that X-ray pictures are totally inadequate for the quantitative estimation of bone mass. Only when as much as 50 percent of bone is lost, do X-ray pictures definitely show these differences. This is actually too late for the early detection of osteoporosis. Single and dual photon absorption apparatus, like a single and double beam spectrophotometer using well-selected mono- or dichromatic gamma rays, measure the bone mass through assessing the absorption. Single photon absorptiometry is used for forearm measurement, and a dual photon method is applicable to the spine and the femoral neck. Reasonably accurate bone measurement by a painless, non-invasive method was accomplished for the first time with this photon absorption technique.

Quantitative computerized tomography (*QCT*) method is another accurate means for an assessment of the spine. By defining the region of interest and using a phantom containing a known amount of potassium phosphate, it is now possible to measure pure *trabecular bone* of the spine. The age-related decrease of the bone is revealed much more clearly with the QCT method than with single or dual photon absorptiometry. When measured by computed tomography, it appears that the soft spongy bone of the spine, in good contact with blood, decreases much faster than the hard cortical bone of arms and legs.

d. Are there different kinds of osteoporosis? Indeed there are. Osteoporosis is a collective term or end result of various processes leading to a decrease in bone mass. Calcium deficiency is only one of these processes. There may be as many kinds of osteoporosis as there are causes of the disease. Two distinct groups are high turnover (rapid rate of both bone resorption and formation) or *active osteoporosis* and low turnover or *inactive osteoporosis*.

Fig. 20 In normal bone, on the left, resorption and formation are in balance. In osteoporotic bone, on the right, more bone is resorbed than is formed, causing a decrease in bone mass. In both normal and osteoporotic bones, the matrix/mineral ratio is kept constant.

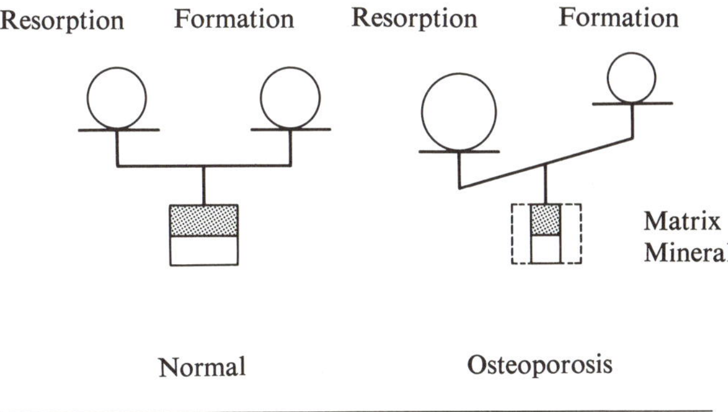

Decreased Bone Mass in Osteoporosis

Fig. 21 High and low turnover osteoporosis.
In both high and low turnover osteoporosis, osteoclastic activity or bone resorption is higher than osteoblastic activity or bone formation. In low turnover osteoporosis, bone resorption is low, but bone formation even lower. In high turnover osteoporosis, bone formation is high, but bone resorption even higher.

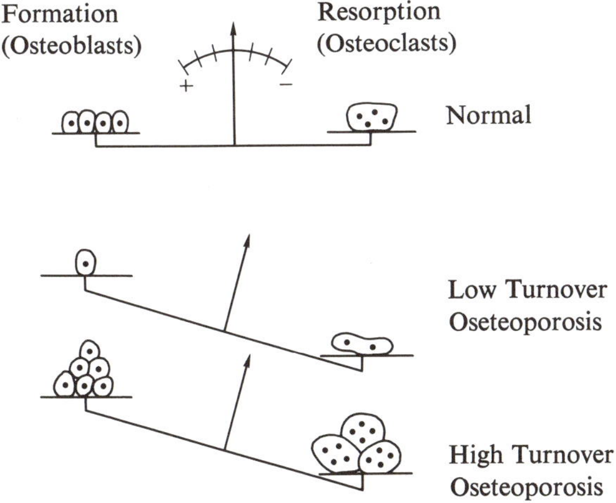

In high turnover osteoporosis, bone is made quite actively. At the same time, bone is resorbed or destroyed even more actively. The result is an overall decrease in bone mass. Sometimes osteoid or uncalcified bone matrix may be increased, because calcification is also quite active (unlike osteomalacia), but just cannot catch up. In this kind of osteoporosis, a slight inhibition of bone resorption may help, and administration of calcitonin and diphosphonate may be useful.

In low turnover osteoporosis, on the other hand, only a small amount of bone is made. Even if bone resorption is also inactive, it nevertheless surpasses the amount which is made, and the result is an overall decreased amount of bone. Low turnover osteoporosis is more difficult to treat than high turnover osteoporosis, because bone formation must be stimulated. Fluoride and parathyroid hormone are possible candidates for the treatment of this disorder, although further studies are necessary before they can be accepted as routine therapeutic agents.

Calcium is the key in the treatment of osteoporosis, but many other agents may also be used in various combinations and time intervals, in order to achieve the best clinical effect. *Bone turnover* (resorption and formation) follows a very slow cycle, taking months to complete. It is therefore quite difficult to evaluate the results of any treatment for osteoporosis. A minimum of two years is usually required before any conclusion can be drawn in a therapeutic trial for osteoporosis.

Since bone is undergoing a slow cyclic remodelling change, attempts are being made to synchronize the drug administration with the natural bone rhythm. Bone activation, depression, free interval (rest period) and repetition (ADFR) are the stages involved in one of the proposed patterns for osteoporosis treatment.

(2) Diabetes and calcium

Diabetes mellitus is a strange disease. Most other diseases give rise to some symptoms. For example, headache and dizziness frequently appear in patients with hypertension. Heartburn, abdominal pain and indigestion accompany gastric ulcer. Diabetes mellitus, however, produces scarcely any symptoms especially in its early stage, except for occasional thirst. When the doctor tells the diabetic patient for the first time that sugar is present in urine, the patient is often surprised, because he does not feel anything is wrong. This is also why the doctor often has a hard time convincing the patient of the need for treatment.

Can we then get along with diabetes doing nothing about it, because it is a mild disease of no consequence? The answer is absolutely No! When diabetes advances, arteriosclerosis develops, involving both large and small blood vessels. When the blood vessels in the eye rupture, blindness may suddenly occur. When the blood vessel supplying blood to the leg is obstructed, necrosis of toes and foot may occur and make amputation necessary. The kidneys may lose their function and the nerves may not properly carry their messages. When atherosclerosis involves coronary arteries, myocardial infarction occurs. Cerebral blood vessels are by no means free from diabetes-stimulated atherosclerosis leading to cerebral thrombosis and paralysis.

Children are sometimes affected by diabetes, but adults and especially elderly people suffer from it much more frequently. Diabetes mellitus is a characteric disease of elderly people. It may be difficult to draw a sharp line between natural consequences of aging and diabetes mellitus, because diseases frequently seen in elderly people, including arteriosclerosis, hypertension and osteoporosis, are often associated with diabetes mellitus.

Diabetes mellitus is caused by deficient insulin secretion. Insulin secretion is decreased under various circumstances. Even if insulin is made to some extent, an obese person may require so much insulin that this amount may become insufficient. The action of insulin may not be fully expressed when other hormones or substances interfere with its action. In any case, a lack of insulin is the major cause of diabetes mellitus.

Insulin is required for glucose entry into cells. This glucose can then be utilized for energy. Not only sugars, but also protein and other nutrients are digested and finally changed to glucose. When the amount of insulin secretion is insufficient, the body cannot utilize glucose, and the glucose level in blood rises. It is therefore necessary to measure the blood glucose level after drinking sugar water, to see if the body can utilize glucose properly. When blood glucose is high, the glucose leaks out into the urine. The presence of sugar in urine is an important and frequently first sign of diabetes mellitus. When the body cannot utilize glucose well, cholesterol and other lipids increase in blood, facilitating the occurrence of arteriosclerosis.

What, then, is the cause of insulin deficiency leading to the rise of the blood sugar level? Various hormones in the body are secreted from many kinds of cells. Insulin is secreted from B-cells of the Langerhans islet of the pancreas. A calcium signal is necessary for the secretion of insulin, as is the case for other hormones.

Our body is composed of about 60 trillion small cells which cannot be seen by the naked eye. Each cell has its own destined function, and completion of its own function is the prerequisite for the health of the whole body. Cells which secrete hormones such as insulin have a special ability to sense the body's need for insulin, such as after a rise of blood sugar following a meal. Calcium is necessary for such sensivity. Insulin secretion does not begin until calcium sends such a signal to the B-cells of the pancreas.

The B-cells of the pancreas are cells specialized to synthesize and secrete insulin. The unique characteristic of these cells is that they can monitor glucose levels. Whenever the glucose concentration in the blood rises, they sense it, and open the calcium gate. The heavy door of the B-cell is thus opened only with two keys, glucose and calcium. When dietary calcium intake is deficient and the inside/outside ratio of calcium across the B-cell membrane is blunted, the calcium signal may not be as effective as before, even if the glucose signal is transmitted normally.

In many elderly diabetics, the pancreas is making enough insulin to maintain normal or even high blood insulin levels. When the body actually requires insulin, however, not enough insulin is available, because of a delay in secretion. Enough insulin does not arrive at the right place at the right time, because the message for insulin secretion is not adequately transmitted. One of the causes of such failure is the disturbance of the calcium message system. Calcium deficiency may blunt the calcium gradient in B-cells of the pancreas. Insulin secretion has been reported to increase after the administration of oral calcium salt, as well as 1α(OH) vitamin D_3, one of the active vitamin Ds.

The action of vitamin D is inseparable from calcium. The active form of vitamin D produced by the kidney is not really a vitamin, but a hormone synthesized and secreted by the kidney from a raw material made in the skin from precursers contained in the food. After it is converted to the active form by the kidney following a preliminary conversion by the liver, it can stimulate calcium absorption from the gut and also help parathyroid hormone to take calcium out of the bone.

The need for the active form of vitamin D in insulin secretion was pointed out by Dr. A. W. Norman and associates at the University of California. Rats fed a vitamin D deficient diet secreted less insulin than those with an adequate vitamin D supply. The addition of vitamin D to the diet completely restored the insulin secretion to normal. Vitamin D evidently facilitated calcium absorption, thereby restoring the calcium signal necessary for insulin secertion.

Most diseases are the results of vicious cycles. Even if something wrong happens in our body, the repair and compensation by other parts of the body are so well designed that restoration is expected within a short time. Disease occurs when one accident causes another one through a chain reaction and this leads to a vicious cycle. In the absence of the active form of vitamin D, insulin production and secretion are markedly reduced. When no insulin is available, it becomes difficult for the kidney to produce the active form of vitamin D. Thus a vicious cycle is formed. In diabetes mellitus, due to insulin deficiency, the active form of vitamin D is also deficient, because of this vicious cycle. If vitamin D is deficient, calcium is poorly absorbed from the gut and the bone gradually loses calcium. This can account for the frequent occurrence of weak, thin bones among diabetics.

(3) Hypertension and calcium

Hypertension is a dreaded disease, representing an important risk factor for cerebral stroke and myocardial infarction. Hypertension and arteriosclerosis are closely associated with each other. Hypertension accelerates the progress of arteriosclerosis, and arteriosclerosis aggravates hypertension. This is again a vicious cycle. Adequate prevention and treatment of hypertension could prolong life by several years.

The causes of hypertension are complex. Apart from the hereditary factor which is rather difficult to control, excessive salt intake has been overemphasized as the cause of hypertension. When hypertension was detected by a doctor, what he would tell the patient at first would be, "Don't take too much salt."

Excessive salt intake certainly appears to be unfavorable for hypertension. On the Japan Sea side of the northeast part of the Japanese mainland, it was customary for people to ingest more than 30 grams of sodium everyday, in forms of extremely salty pickles, salty soup and salted fish. Through the long winter, these were the only foods available. Salt is the oldest food preservative known to mankind. Surprisingly high prevalence rates of hypertension and cerebral apoplexy were found in these districts. With the help of public education, this excessive salt intake was eventually reduced, and the incidence of cerebral apoplexy also decreased.

Sodium might not be the only culprit in these areas. On the Pacific side of the northeast district, across a line of mountains, people have taken as much salt but there is less snow and more sunshine in winter than on the Japan Sea side. On the Pacific side, the prevalence rates of hypertension and cerebral apoplexy have been much lower than those on the Japan Sea side. In these areas, dietary sodium excess was similar, but the people enjoyed more sunshine throughout the year than on the Japan Sea side. Insufficient sunshine exposure could lead to inadequate vitamin D synthesis in the skin and consequently a decrease of gut calcium absorption. In addition to excess sodium intake, deficient calcium intake may play a role in the development of hypertension, perhaps even a more decisive one.

Dr. David McCarron, of the University of Oregon, recently made a very important observation. According to the results of dietary surveys, people suffering from hypertension consumed less calcium than people with normal blood pressure, but sodium intake was not different between the two groups (McCarron, 1982). The results suggested that calcium deficiency plays a more important role than sodium excess in the development of hypertension. Dr. McCarron has presented his "calcium theory of hypertension" on television. When he discusses his point, he always meets firm opposition. There are yet too many who adhere to the traditional "salt theory of hypertension." Since this problem is so important for everyone and so directly concerns our daily food, it is desirable to discuss it fully and draw a conclusion.

Calcium deficiency stimulates parathyroid hormone secretion and this causes calcium to be withdrawn from the bones, so that the calcium enters the blood vessel walls. Calcium in the blood vessel wall causes contraction of the blood vessel, leading to hypertension.

How is high blood pressure produced? Hypertension is like too much water running through the water transport system, exerting a high pressure against the walls. The wall of the water pipe is pushed by a great strength. The heart as a pump is primarily responsible for such power. People with high blood pressure almost always have a big *hypertrophic* heart. Blood pressure may be high because the heart is too strong, or the blood vessels are too narrow, and the resistance to the blood stream too high because of the contraction of the blood vessel.

Blood pressure may rise when too much blood is circulating. The amount of blood circulating through our body is said to correspond to about 1/13 the weight of the body. If a person weighs 65 kg, about 5 kg or 5 kl blood are probably circulating through the body. Blood is pumped out of the heart and passes through the aorta to smaller and smaller arteries until finally arriving at the capillaries. After passing through the tissue, it returns to the heart via the veins. If the amount of blood increased to 6 kl, the vascular tree would be overcrowded with blood and an increased pressure exerted on the blood vessel wall. This would result in hypertension induced by an increased blood volume.

When too much salt enters the body, the blood becomes too salty and this may draw water out of the cell to keep a constant taste. In order to avoid such an extreme state and to provide a safe salt concentration, water enters the blood to dilute it. As the water enters the blood, the volume of blood increases. Excessive salt intake thus increases the blood volume causing hypertension.

However, even if the body is flooded with salt and blood volume is increased, attempts are always made to restore the normal metabolism by increased sodium excretion. Thus chronic hypertension does not easily develop from sodium excess alone.

What, then, is the true cause of hypertension in most of the patients? The contraction of small blood vessels decreases the volume within the blood vessel. Even if the overall blood volume remains the same, blood pressure rises on such vascular contraction.

Adrenaline and *noradrenaline*, hormones secreted from the *adrenal medulla*, raise blood pressure by contracting blood vessels. One of the actions of these hormones is to increase the calcium inflow to the smooth muscle cells of the blood vessel, strengthening the contraction of the blood vessel.

When calcium is deficient in the diet and parathyroid hormone takes calcium out of the bone, the calcium enters the cells of the blood vessel wall to contract the smooth muscles, causing a narrowing of the blood vessels. This would increase the resistance and raise the blood pressure.

Contraction of small blood vessels all over the body thus causes hypertension, and hypertension decreases the oxygen supply to the cells of the vascular wall, increasing calcium entry and contraction of blood vessels in a vicious cycle. Stiff shoulders and headache are frequently associated with hypertension, probably as the result of such a process.

The intimate relationship between calcium and hypertension was further confirmed by experiments in spontaneously hypertensive rats (McCarron et al., 1985). Professor Kozo Okamoto of Kyoto University mated rats with hereditary hypertension and was able to produce rats with spontaneous hypertension. The hereditary high blood pressure in these rats is quite similar to the high blood pressure in humans. This is one of the best and most convenient models for studying human hypertension and to examine the effect of drugs on hypertension. The mechanisms of disturbances in calcium metabolism in these rats are now being studied.

In these spontaneously hypertensive rats, the metabolism and function of vitamin D appears to be abnormal, thereby causing abnormal calcium absorption from the gut. When abundant calcium was given to these rats, the blood pressure stayed low despite the hereditary tendency to become hypertensive. When the amount of calcium in the food was decreased, the blood pressure rose rapidly. When calcium intake was low, more parathyroid hormone was probably secreted, causing calcium to be drawn from bone and deposited in the blood vessels. As a result, blood vessels contracted, raising the blood pressure.

When parathyroid glands were removed from these rats, even a calcium deficiency failed to cause hypertension. Since *parathyroidectomy* interrupted this chain of events—calcium deficiency, parathyroid hormone secretion, calcium mobilization from bone, calcium deposition in the blood vessel—blood pressure did not rise.

The contraction of the blood vessel caused by the entrance of calcium into the vascular wall is the final common pathway in the development of hypertension due to various causes. The mechanism of hypertension may now be explained much more clearly by this mechanism.

Even when hypertension is caused by excessive salt intake, calcium appears to play a definite role. The calcium concentration outside of the cell is quite high, 10,000 times that inside of the cell. The cell membrane prevents outside calcium from coming in, similar to the walls of the castle. Like the old castle gate where the watchman scrutinized everyone who arrived and permitted only authorized persons to enter, the cell membrane also has a calcium gate. This gate allows nothing but calcium to enter. It is the gate which keeps excess calcium outside the cell, to keep the cell alive and active.

There is another mechanism of communication for calcium exchange between inside and outside of the cell, as prisoners of war are sometimes exchanged between two countries during wartime. When sodium goes out of the cell, calcium is brought into the cell in its place. The entry of sodium into the cell is not as severely restricted as that of calcium, so sodium enters the cell rather abundantly when sodium is abundant outside. When too much salt is ingested, sodium thus simply enters the cell, since no Sodium Paradox operates, unlike calcium metabolism. When the cell has abundant sodium, the sodium inside the cell is exchanged with the calcium outside the cell. The result is an increase of calcium inside the cell. *Aldosterone* and *deoxycorticosterone*, so-called salt retaining hormones, decrease the excretion of sodium in urine, causing flooding of sodium in blood.

The inside of the cell is also flooded with calcium eventually. Calcium then enters the cell in exchange for sodium.

Hypertension caused by excess sodium intake may also be explained by this increase of calcium inside the cell.

Is it possible to prevent and treat hypertension after all? Various methods have been suggested. Avoidance of overwork and strain is important. Many drugs are available to lower the blood pressure effectively. The value of salt restriction in the treatment and prevention of hypertension has been questioned, unless an extraordinarily large amount of salt is taken, as in the northeastern part of Japan.

When one becomes too anxious about decreasing sodium intake, one might also inadvertently decrease the intake of calcium. Even if the sodium intake is successfully restricted, simultaneous calcium deficiency might be even more harmful. Sufficient calcium intake along with moderate salt intake would appear to be the safest and most effective method. For the prevention of hypertension, the importance of sunbathing and sufficient vitamin D and calcium intake should not be underestimated.

Calcium antagonists are a group of drugs preventing the entry of calcium inside the cell. When this type of drug is used, the gate of the cell wall is tightly locked to prevent calcium entry, and blood pressure rapidly drops.

Sufficient intake of dietary calcium as well as calcium antagonists prevent calcium from entering the blood vessel wall. Calcium thus acts as a calcium antagonist. In fact, calcium is the best calcium antagonist. This is another example of the Calcium Paradox.

When sufficient calcium is given orally, it prevents parathyroid hormone overactivity and increased calcium removal from bone, thereby inhibiting entrance of calcium into blood vessel walls. In this sense, calcium is a calcium antagonist, capable of lowering the blood pressure. When the same calcium is injected intravenously, calcium raises blood pressure.

Magnesium is also a calcium antagonist, preventing entry of the calcium into the cell. It is no wonder, then, that magnesium deficiency has also be shown to raise blood pressure. The action of magnesium is probably mediated by calcium.

(4) Arteriosclerosis, myocardial infarction and calcium

Arteriosclerosis is a hardening of the arteries. The artery is originally as soft and elastic as rubber. The smooth muscle incorporated into the wall contracts and relaxes according to the needs of the body, changing the size of the artery to adjust to the amount of blood passing through.

When the artery hardens, first, free contraction and relaxation become impossible. Second, the inside of the hardened artery becomes narrower, permitting less blood to pass through. The hardened blood vessel wall is then exposed to higher pressure. If some part of the wall is weak, this part of the wall may protrude like a sac, or may even rupture. This is called an *aneurysm.*

As one becomes older, such changes progress. Why does the wall of the artery become hard? When such hardened arteries are examined, large amounts of calcium are found inside. The hardened artery is also thickened and the inside of the artery becomes narrower. Some blood cells can become trapped inside, with the *lumen*, or open space with an artery, becoming even narrower. When fibers in the vascular walls are increased and smooth muscle cells are decreased, the arterial wall becomes quite rigid. One can even break the really hardened artery like a piece of wood. Calcium deposition is one of the major causes of such extreme hardening.

Just as excessive salt intake has taken the blame for hypertension, excessive cholesterol is the scapegoat for arteriosclerosis. Atherosclerosis, or hardening of the arterial wall from the inside because of the formation of lipid plaques, is intimately related to cholesterol. The relationship between cholesterol and arteriosclerosis began to attract attention in the 1940s and 1950s. When the food supply became abundant after World War II, people become overweight from the excessive intake of fat and calories, as well as from inadequate exercise. Mental stress accelerated these changes, by encouraging overeating, possibly for tension relief. Myocardial infarction became very frequent around this time, especially in middle-aged and younger subjects. When serum cholesterol was measured, it was almost invariably high in these patients suffering from premature myocardial infarction.

Myocardial infarction is caused by arteriosclerosis and consequent obstruction of the coronary artery supplying blood to the heart muscle. When blood cannot pass through this part of the coronary artery, the part of the heart muscle which used to receive blood from this artery can no longer stay alive. In the heart and also in the brain, one portion of the heart muscle receives blood only from one artery, not like the liver where a network of blood vessels brings blood from everywhere. A loss of blood supply from one artery therefore means death to the part of the heart muscle receiving oxygen and nutrients through this artery.

The final process of cell death is related to calcium. As the cell becomes deficient in oxygen, the cell membrane loses control over the calcium entrance from outside, resulting in a flow of calcium into the cell. This results in destruction of protein and loss of electric activity.

When a large part of the heart muscle is damaged and cannot contract any more, blood cannot be delivered to vital organs such as the brain. The blood pressure falls causing a state of shock, and nothing can save the patient if the heart stops beating for a few minutes. When the same arterial obstruction occurs in the brain, cerebral infarction occurs, leading to paralysis, speech disturbance and other difficulties.

When the portion of heart muscle destroyed from loss of blood supply is small, the heart somehow manages to continue pumping the blood, but its strength is reduced. Heart failure or irregular rhythm may occur, and the patient may be incapacitated.

The human body is a mystery. Whenever something dangerous or harmful comes into the body, a variety of efforts are made to eliminate it or change it to a harmless material in order to maintain health. The inside of a mammoth tanker carrying 200,000 tons of oil is divided into numerous small chambers. Even if one chamber catches fire, the protecting wall between each chamber is so efficient that the fire is confined to one chamber and does not spread all over to cause a disastrous explosion. Our body is also meticulously compartmentalized to prevent infections from spreading.

The protecting wall of the blood vessel is called the *internal elastic plate*. This separates the inner layer of the blood vessel adjacent to the blood stream from the *media* of the blood vessel, which occupies the main part of the vascular wall and contains smooth muscle. In addition to giving elasticity to the blood vessel, the internal elastic plate prevents the blood vessel components, such as cholesterol, from entering the main part of the blood vessel. The internal elastic plate is thus the defense corps along the border. As long as the internal elastic plate is intact, even a high level of cholesterol in blood does not readily enter the blood vessel wall.

Even the strongest defensive wall cannot be without some weak point. Remember the story of Troy. The hosts of the Greek army surrounded the wall of the city of Troy and continued to attack it for many years. The Trojan army also fought well under the command of Achilles and stood firmly. The trick devised by the Greeks was a huge wooden horse which could hold many soldiers. When the Greeks retreated leaving the wooden horse with the soldiers, the Trojans thought they had won, and pulled the wooden horse inside the city wall. While the Trojans were sleeping after their victory celebration, the soldiers came out of the horse and broke the gate of the city wall to allow the Greek army inside.

Calcium is the Trojan horse and cholesterol is the Greek army. Calcium at first breaks the internal elastic plate to open the way for cholesterol to invade the blood vessel wall. In addition, high blood cholesterol levels enhance calcium entry into the cell, aggravating the situation.

The calcium content of the blood vessel increases with age, beginning from early childhood (Lansing et al., 1950). When blood vessels were analyzed, calcium was found to be associated with *elastin*, the main component of the internal elastic plate. This calcium-elastin

Fig. 22 Increase of calcium in the blood vessel with age. The increase begins in childhood (A. I. Lancing, 1950).

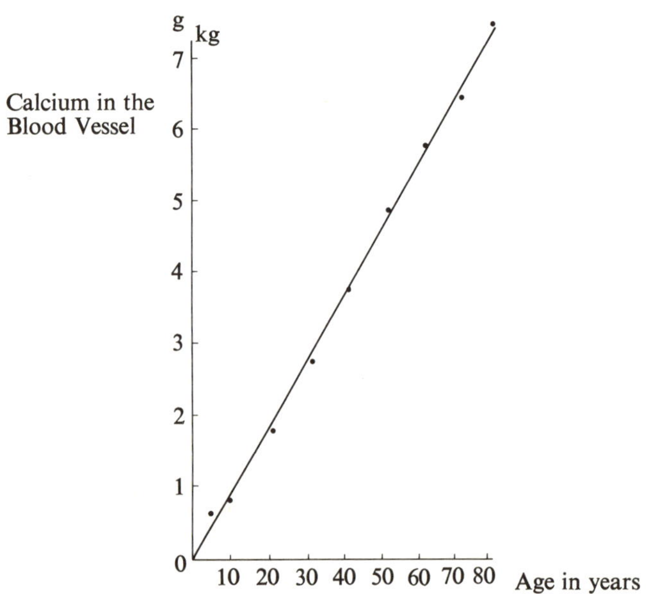

complex had lost the original elasticity of elastin. Since the blood is exposed to constantly changing blood pressure, such rigid "calcium-elastin" readily breaks in response to the stretching from blood pressure changes. No wonder the firm defensive line of the internal elastic plate is broken, thus permiting the cholesterol into the blood vessel wall.

Calcium, and not cholesterol, thus pulls the first trigger for the development of arteriosclerosis. Cholesterol of course follows calcium and continues the process. Calcium may therefore be called the *initiator*, and cholesterol the *promotor* of arteriosclerosis. When a large amount of cholesterol is finally deposited in the blood vessel wall and the process of arteriosclerosis is almost complete, calcium is heavily deposited in the blood vessel wall, making the artery visible in the X-ray picture. The real hardness of the *sclerotic artery* is given by the calcium deposition. Arteriosclerosis is said to begin with calcium and end with calcium. Even before the artery shows any changes of arteriosclerosis or cholesterol content, the calcium content of the blood vessel wall has already increased.

A man is as old as his arteries. Even if the heart and brain are quite young themselves, the aging and arteriosclerosis of the arteries supplying blood to these organs would cause a loss of their function. Since calcium in the artery is really responsible for the appearance of arteriosclerosis, a man is as old as the amount of calcium in his blood vessels. Sufficient intake of calcium prevents calcium from entering into the vascular wall, thus keeping the arteries young. Thus calcium in food, not the calcium in the artery, may keep us young.

Arteriosclerosis and hypertension go hand in hand. Many people have both of these diseases. Hypertension predisposes the arterial wall to injury, especially in the elastic plate, causing a loss of the elasticity of the blood vessel. This permits cholesterol and other lipids to enter the vascular wall and be deposited. Arteriosclerosis also facilitates the occurrence of hypertension, through the loss of elasticity of the blood vessel. Such a rigid artery cannot properly meet the push by blood pumped out of the heart, and therefore the mechanical stress cannot be attenuated. Decreased supply of blood to the kidney stimulates the release of substances which further raise the blood pressure. Thus hypertension aggravates arteriosclerosis, and arteriosclerosis aggravates hypertension in a vicious cycle. Calcium deficiency is a major cause of both hypertension and arteriosclerosis. Adequate calcium intake could help to prevent both of these dreadful diseases, interrupting the destructive cycle.

When you travel to Europe, you are probably advised not to drink tap water because of poor taste. Soil calcium content is very high in Europe, so that water also contains a large amount of calcium. Fruits and vegetables also reflect the calcium content of water. Drinking water is sold in bottles. In only a few countries in the world, including the United States, is water free and safe to drink. In many places in the world, hard water is supplied as tap water. The discovery of calcium in England was based on a spontaneous precipitation of calcium from hard water as white crystal. Hard water containing abundant calcium is hard to drink, but it may also have some advantage. Many epidemiological studies have suggested that people living in hard water areas suffer from arteriosclerosis less frequently than people living in soft water areas (Crawford, 1972). When rain falls on earth, it is almost pure water containing no minerals, like distilled water

Fig. 23 Calcium content of the soil is different among various countries. It is generally high in Europe.

Calcium Content of the Soil (%)

Country	Calcium Content of the Soil (%)
Japan	0.63
U.K.	3.83
France	4.10
Germany	1.35
U.S.A.	1.31

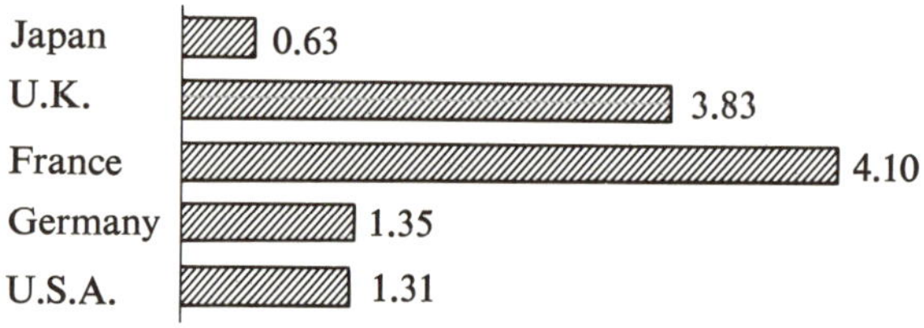

Fig. 24 Calcium content of fruits and vegetables in Europe and Japan (mg/100 grams).
Most of the fruits and vegetables in Europe contain more calcium than those in Japan.

Food	Japan	Europe
Turnip	25.0	58.7
Cucumber	19.0	22.8
Cabbage	45.0	53.2
Potato	5.0	7.7
Tomato	3.0	13.3
Leek	40.0	62.7
Carrot	35.0	48.0
Fig	29.0	34.2
Strawberry	14.0	22.0
Cherry	10.0	15.9
Plum	6.0	13.7
Peach	3.0	4.8

made in the laboratory. As the water runs through the rocks and soil, minerals including calcium are mixed in to make delicious water. Thus the higher calcium intake may decrease the incidence of atherosclerotic disease like coronary artery disease (Knox, 1973). The calcium content of the coronary artery is apparently lower when calcium intake is higher. When sufficient calcium is ingested, serum lipid, including cholesterol, decreases, possibly because calcium combines with fat to interfere with its absorption (Yocowitz et al., 1965).

(5) Calcium and liver disease

There are many kinds of liver disease. *Serum hepatitis*, caused by a virus is transmitted through blood transfusions. Liver cirrhosis and even *hepatoma*, cancer of the liver, are among the possible consequences of hepatitis. People drinking alcohol frequently suffer from hepatitis and fatty liver. Drugs and chemicals may also cause liver damage. *Carbon tetrachloride* is one of the most toxic substances for the liver.

What happens when the liver cell is damaged? At first, the function of the cell membrane surrounding the cell is impaired, and calcium comes into the cell from outside. When a large amount of carbon tetrachloride gas is accidentally inhaled, the liver is damaged so severely that the color of the liver changes from normal reddish brown to yellow-white. The calcium content of each liver cell is increased tremendously. What really happened was that the cell membrane was damaged by the chemical, causing entry of calcium into the cell and subsequent liver damage.

Also in other diseases of the liver, or liver cell damage, calcium always enters the inside of the cell. A cell becomes sick when too much calcium enters the cell, decreasing the inside/outside difference of the calcium. The liver cell is no exception. When a virus attacks the cell or alcohol damages the cell, calcium inside the cell always increases.

The progression of liver disease is intimately related to the immune function. The virus damages only a small number of liver cells, certainly not all the liver cells. When the immune function of the body is too strong or directed to a wrong target, the immune mechanism attacks such cells as if they were invaders from outside. This attack is directed to other liver cells as well. Such indiscriminate attack is like a trigger-happy policeman who shoots anybody with any similarity to the criminal. Such over-excitability of the immune system might be hereditary.

Certain patients who suffer from *viral hepatitis* show a rapid progress of the disease leading to chronic hepatitis, liver cirrhosis and hepatoma. Insufficient treatment might represent one of the factors, but a hereditary disposition to indiscriminate excitability of the immune system might also be a contributing factor. Diseases of the liver progress in two stages, the first stage based on a direct damage of the liver cell by virus, chemical substance or others, followed by the second stage consisting of autoimmune damage of the injured liver cells, and other nearby liver cells.

Some people drink a large amount of alcohol but display no liver damage, while in other people a relatively small amount of alcohol may cause profound damage to the liver. Such differences are based in part on differences in hereditary constitution, especially in the mode of the immune response. Alcohol always damages a small proportion of liver cells. In people with overactive and indiscriminate immune function, the remaining innocent liver cells are destroyed, while no such disaster occurs in other, more fortunate, people.

Would it be possible to prevent such undesirable and dangerous liver damage? Although a hereditary disposition is extremely difficult to change, influence on the immune response may be possible by the use of drugs called *immunoregulators*.

Sufficient calcium intake and the active form of vitamin D may act as immunoregulators. Whenever a cell is damaged, liver cells or others, there is always an increase of calcium inside of the cell. Since calcium deficiency tends to increase calcium inside of the cell through the action of parathyroid hormone, calcium supplements may prevent this tendency and reverse the liver damage caused by virus, toxic substances including alcohol, and indiscriminate immune function. Sufficient calcium intake may be of value in preventing the aggravation of liver diseases. Calcium itself may also be a good immunoregulator. Among various kinds of lymphocytes forming an immune network, suppressor cells are responsible for calming down the overzealous "trigger-happy" lymphocytes. Adequate nutritional calcium intake, providing optimal inside/outside calcium ratio in these suppressor lymphocytes, would ensure excellent discipline in the teamwork among lymphocytes, precluding indiscriminate attacks to our body.

(6) Overweight and calcium

Overweight is first of all the result of overeating. If you keep eating too much, especially sweet things, the body weight will increase. As you gain weight, the amount of food you used to take when you were lean does not satisfy you any more. The more you eat, the higher the blood sugar rises and the more insulin is secreted, causing the blood sugar to fall. This naturally makes you more hungry, and a vicious cycle is initiated. Overweight thus knows no limit unless it is drastically stopped somehow; in extreme cases, a large portion of bowel is cut to decrease the absorption of nutrients from food. Most of the weight gained by excessive caloric intake is based on the increase of fat. Excessive fat in the body is obesity, which seriously interferes with the actions of the heart and lung.

Obesity can also be caused by excessive secretion of hormones controlling sugar metabolism called glucocorticoids, produced in the adrenal cortex. This condition is called Cushing's syndrome, after the famous neurosurgeon who first described it. Deficient thyroid hormone may also cause obesity. Many people with diabetes are obese, but this may be the cause and not the result of diabetes. Women tend to gain weight especially after middle age.

Why do people eat too much? In most cases, a psychological imbalance or conflict is present. When we are not happy with people around us, or we are nervous and anxious, we tend to eat more. Peace of mind is one of the effective measures against overeating. A strong will to refrain from overeating also requires peace of mind. When the calcium level in blood is low, one could become nervous and anxious. Calcium may help these people to stabilize the emotional state. Patients with *pseudohypoparathyroidism*, with low blood calcium because the cells cannot respond to parathyroid hormone, usually have a round face, and are obese. Subtle decreases of calcium in blood causing excessive parathyroid hormone secretion may be one of the causes of obesity.

Calcitonin is a hormone which is secreted when blood calcium is high and calcium is sufficiently taken. It is not secreted when calcium is deficient or when blood calcium is low. In addition to bringing calcium from the blood back into the bone, calcitonin controls and suppresses appetite. When calcium is deficient and calcitonin secretion decreases, it becomes more difficult to control the appetite. Calcitonin secretion is generally lower in females than in males, and becomes even lower after middle age. Increase of appetite and overweight especially among middle-aged women may be explained in part by calcitonin deficiency. Enough dietary calcium may increase calcitonin secretion and control appetite.

A deficiency of vitamin D may also be related to obesity. Experiments have been conducted on a particular strain of rats which have an insatiable appetite and become obese soon after birth, and which invariably develop diabetes mellitus. When the active form of vitamin D was given to these rats, the rats lost their appetite, lost weight and diabetes mellitus also improved. This may have been due to the action of active vitamin D in increasing calcium absorption from the gut and increasing blood calcium levels. Calcitonin may also have been secreted in response to the calcium excess, decreasing appetite and curing both the obesity and diabetes mellitus.

It is suggested that calcium combines with fats contained in the food to produce unabsorbable calcium soaps. Calcium supplements have been reported to lower the level of cholesterol and other fats in plasma by preventing their absorption.

Overweight and obesity can cause various diseases of old age. Therefore, weight reduction is important for people who are overweight. When food is restricted to reduce weight, one should not do so too rapidly or fanatically. Adequate calcium intake should be maintained even when food is restricted.

(7) Peptic ulcer and calcium

Peptic ulcer is another disease of civilization, or the product of stress. Excessive mental or physical stress causes secretion of a large amount of steroid hormone from the adrenal cortex, which increases the secretion of gastric juice. Too much gastric juice digests the gastric wall when food is not around. This is the main cause of peptic ulcer which may occur in the stomach and duodenum. In addition to neutralizing acidic gastric juice, calcium may stimulate calcitonin secretion, and calcitonin inhibits gastric secretion, healing peptic ulcer. Emotional stabilization is also effective for relief of peptic ulcer.

(8) Stones and calcium

Stones are sometimes formed in the kidney and gallbladder. Even if they are small, they may cause a lot of trouble. Pains do not stop until the stones are removed. The stream of urine may be obstructed by kidney stones, while bile flow may be stopped by gallstones. Stones may also injure the walls of these organs causing bleeding or infection. Stones may also be formed in the pancreas or salivary glands. Not only minerals like calcium and phosphorus, but also organic materials such as uric acid and *cysteine* form kidney stones. Cholesterol is a major component of gallstones. Since most of the stones contain some calcium, people frequently worry about stone formation when taking supplemental calcium.

Kidney stones are the most common form of stones. Calcium is usually dissolved in blood and also in urine and should not be precipitated at places other than the bone. Under exceptional circumstances, calcium in urine may precipitate to form stones. First, a large amount of calcium facilitates stone formation. Second, alkaline urine facilitates calcium precipitation. On the other hand, acidic urine facilitates formation of uric acid and cysteine stones.

Many patients with calcium kidney stones excrete a large amount of calcium in urine which is frequently alkaline. Calcium taken as food is by no means directly excreted in urine. People suffering from kidney stones are not necessarily taking a large amount of calcium in food.

When we eat something, it passes through the gut and comes out in the stool unless it is absorbed from the gut wall. Even if something like a coin is accidentally swallowed it usually comes out in the stool safely, confirming the point that inside the gut is still outside of the body.

Gut mucosa act like a strict immigration control. Those nutrients necessary for the body are absorbed, but materials not required by the body are not absorbed, and merely allowed to pass through.

A large amount of calcium is excreted in urine when abundant calcium is coming out of the bone. This might be called kidney stone coming out of the bone. Bone is like a safety deposit of calcium. In general, unless the key for the deposit, parathyroid hormone, is present, calcium does not come out of the bone. When parathyroid hormone is secreted in excess, in a disease called primary hyperparathyroidism, a large amount of calcium comes out of the bone and floods the kidney, causing kidney stones to form.

When calcium intake is insufficient, parathyroid hormone causes calcium withdrawal from the bone to prevent a decrease of blood calcium. More calcium may be secreted in the urine this way, prompting the formation of a kidney stone. On such occasions, increase of dietary calcium intake might paradoxically prevent kidney stone formation.

Some people absorb calcium so well from the gut that they excrete large amounts of calcium in urine to form kidney stones even if the calcium in food is adequate and not too much calcium is coming out of the bone. This is called absorptive type hypercalciuria. In this condition, kidney stones are formed from calcium coming from the gut. The reason these people absorb a large amount of calcium from the gut is not yet fully understood. When a large amount of vitamin D is taken, similar augmentation of gut calcium absorption occurs. People with absorptive hypercalciuria may be producing too much vitamin D. For such people, the more dietary calcium they ingest, the more calcium they absorb from the gut and excrete in the urine to form kidney stones. For most others, the intestine efficiently controls calcium absorption. When too much calcium is taken as food, it simply is not absorbed.

In order to decide the type of kidney stone, certain measurements are necessary, including calcium in blood and urine tested under varying levels of dietary calcium intake. If calcium excretion in urine increases with high calcium intake and decreases upon calcium restriction, the patient probably has absorptive type hypercalciuria since urinary calcium excretion depends on food calcium content and intestinal calcium absorption. When calcium leaks through the kidney, changes in calcium intake do not change urinary calcium excretion. This is called renal hypercalciuria or kidney stone due to renal calcium leak. Thiazide drugs which lower blood pressure also decrease urinary calcium excretion, and help prevent kidney stones.

(9) Cancer and calcium

Cancer is the number one enemy of human beings, developing within our own bodies and threatening our lives. Children may sometimes suffer from cancer, but a large proportion of cancer occurs after middle age. The cause of cancer is unknown, but undoubtedly quite complex.

Calcium is the flame of life, required in all cell functions including cell division and proliferation. When no calcium is available, cells do not multiply or grow. The calcium signal is essential for cell multiplication. The trouble with the cancer cell is that it never stops multiplying once it starts. It is only natural to suspect that in cancer, something is wrong with the calcium signal.

Parathyroid hormone is secreted whenever blood calcium is low and activates calcium withdrawal from bone. When too much parathyroid hormone is secreted, it might take too much calcium away from the bone even if it is not necessary. This is a disease called primary hyperparathyroidism. Bone loses calcium endlessly and blood calcium rises. The excess calcium goes out in urine through the kidney to make stones.

Dr. S. G. Massry and his group at the University of Southern California compared the frequency of cancer among 100 patients with primary hyperparathyroidism and those with other diseases (Kaplan et al., 1971). Patients with primary hyperparathyroidism surprisingly had more cancer than patients of similar ages with other diseases;

the cancers included those of the thyroid, breast, stomach, gallbladder and other organs. Something must have been stimulating the cells to become cancerous when too much parathyroid hormone was being made. Parathyroid hormone secretion causes calcium removal from bone and also increases calcium in many cells. Whenever calcium increases within the cell, growth and proliferation of the cell are stimulated. The development of cancer is a complex process involving many factors. An increase of calcium in the cell appears to be one of the factors favoring the development of cancer. This is one reason why surgery should be promptly carried out to treat primary hyperparathyroidism, even if symptoms are mild or absent.

The increased number of people suffering from cancer in advancing age may be explained by the increasing degree of calcium deficiency with progressive activation of parathyroid hormone. Patients with renal failure who are on dialysis also suffer from cancer more frequently than other people, possibly because of calcium deficiency and parathyroid overactivity, as in elderly people, because the kidney is no thyroid overactivity, as in elderly people, because the kidney is no longer able to produce the active form of vitamin D necessary for intestinal calcium absorption. In the absence of calcium and vitamin D, nothing prevents the parathyroids from becoming overactive.

Gastric cancer is quite common in Japan, in contrast to the United States where this form of cancer is rare. Many Japanese people take much less calcium than most American people and seldom drink milk. When the frequency of gastric cancer was compared between the milk-drinking and non-milk-drinking Japanese population, gastric cancer was found to appear much less frequently in the milk-drinking group. By drinking a pint of milk daily, it is possible to reduce the risk of gastric cancer to one-half. Even in Japan, the incidence of gastric cancer tends to decrease as people start to drink more milk. In the United States, colon cancer was recently reported to occur much less frequently in people taking enough calcium and vitamin D than in those who are taking less.

When a part of the liver of a rat is removed, the remaining liver suddenly starts to grow to make up for the loss. This is called *liver regeneration*. This is like regrowth of the tail of a lizard. After removal of the parathyroid glands, the speed of liver regeneration becomes much slower (Rixon, 1972). Parathyroid hormone is thus important in stimulating the growth and proliferation of liver cells. Calcium gets removed from the bone, and an increase of calcium in the blood and cells probably promotes liver cell growth and multiplication.

In the 1960s, when the threat of all-out nuclear war was emphasized for the first time, research efforts were concentrated on searching for drugs or substances to protect the human body from the dreadful effect of radiation, so-called *radioprotective agents*. For the group of Drs. R. H. Rixon and Whitfield in Ottawa, it must have been quite an unexpected finding that parathyroid hormone showed a potent radioprotective effect (Rixon et al., 1961). When parathyroid extract was injected into rats, the rats tolerated radiation much better and survived far longer than the controls without any parathyroid hormone. The effects of radiation are mainly manifested in rapidly growing and multiplying cells, such as the cells in the bone marrow, where blood is formed, and cells lining the gut wall. Two major fatal consequences of radiation are losses of bone marrow cells and intestinal epithelial cells leading to a decrease of all blood cells and hemorrhage from the gut. Since parathyroid hormone mobilizes calcium from blood and increases the flow of calcium from outside to inside the cell, parathyroid hormone helps the bone marrow cells and gut surface cells to grow and multiply in order to recover from the injury caused by the radiation.

Too much parathyroid hormone does harm to the human body, as in the disease primary hyperparathyroidism. Increase of calcium inside the cell and blunting of the outside/inside concentration difference of calcium impairs the function of the cells involved in the immune network. Since one of the important functions of the body's immune network is to attack and control cancer cell growth, a decreased immune function might give cancer cells an opportunity to grow, just as the occurrence of robberies may increase when policemen are absent. Cancer cells are originally cells of our own body, but have been transformed to become uncontrollable by multiplying endlessly, taking nutrients from the body, and destroying neighboring tissue. The immune system should be able to recognize cancer cells as something unusual and harmful and should try to remove these cells. The function of this so-called immune surveillance system is to watch all over the body to check if something is wrong and to take care of anything which is unusual and different from our own cells. Calcium deficiency and blunting of the outside/inside ratio of calcium around the cell causes a disorder and confusion of this immune surveillance system. It is quite possible that cancer cells take advantage of this condition and start to multiply.

When we think of cancer, we always think of something quite dreadful. Nothing appears to stop its growth. Actually, many cancer cells may be appearing in our body from time to time. Because of the immune surveillance system, in healthy people, these cancer cells appearing in the body are killed before developing into full-fledged cancer. If there is any trouble in the immune surveillance system, cancer cells are more likely to develop into real cancer. One of the dreadful consequences of calcium deficiency is the disturbance of the immune surveillance system.

One of the phenomena frequently found in cancer patients is the rise of blood calcium. Most of the calcium in blood has come out of the bone. In a large hospital, blood calcium is measured in thousands of patients every day. What disease raises blood calcium level? It is cancer that most frequently raises the blood calcium level. Blood calcium is also high when too much parathyroid hormone is secreted, as in primary hyperparathyroidism, or when too much vitamin D is taken by mistake, causing vitamin D intoxication. Cancer may release some substance acting like parathyroid hormone or vitamin D, to cause calcium to be withdrawn from the bone and to increase calcium in blood. When the cancer is successfully removed from such patients, blood calcium may become normal. When calcium in the blood is too high, appetite decreases, muscle power falls and the patient becomes less active mentally. At times, consciousness may be lost. Cancer patients become very ill, partly because blood calcium is too high. When calcitonin or other drugs successfully bring the high blood calcium down, the patient may become stronger and live longer. The significance of high blood calcium in cancer patients remains unknown, but it may represent a desperate attempt of the patient to control cancer, by suppressing parathyroid hormone secretion which could promote cancer growth.

Drugs called *calcium antagonists* prevent calcium from entering the cell. When anticancer drugs are used along with calcium antagonists, their action is augmented. Decrease of calcium entry into the cell thus prevents cancer cells from growing.

As previously mentioned, the active form of vitamin D stimulates calcium absorption from the gut. Calcium binding protein is rapidly synthesized in the gut surface cells by the action of vitamin D. Since a very large amount of calcium must pass through the gut cell before entering the blood, the increase of calcium within the cell may damage the gut cell. This could be modified, similar to a process of detoxification, if the calcium binding protein binds the calcium to decrease the free calcium within the cell.

This action of vitamin D is in detoxifying the cell of excessive calcium. This action may protect many cells from the loss of function by the entrance of too much calcium into the cell.

Leukemia is cancer of blood cells. Leukemia cells, like cancer cells, continue to multiply until they devastate bone marrow, causing the disappearance of normal red cells, white cells and platelets. Dr. T. Suda of Showa University, Tokyo, was studying leukemia cells cultured in a test tube. While the effects of many drugs on leukemia cells were tested, the active form of vitamin D was shown to profoundly change the property of leukemia cells. Active vitamin D actually changed wild uncontrollable leukemia cells into tame and even useful macrophages (Abe et al., 1981). If such an effect were achieved within the human body, a valuable addition would be made in our defense armory against leukemia. Unfortunately, the effect of the active form of vitamin D on human leukemia has not yet been conclusively demonstrated in patients. The reason why the active form of vitamin D had such a dramatic effect on leukemia cells has not yet been fully explained.

In any case, calcium is a truly mysterious substance which changes the course of cell growth, multiplication, and differentiation. Calcium is one key to understanding and controlling cancer cells.

(10) Alcohol and calcium

Drinking too much alcohol is naturally harmful. The problem is "How much is too much?" People drink alcohol in different ways. Alcohol has a toxic effect on the liver, and often contributes to nutritional deficiencies of protein, vitamins and calcium. This often causes metabolic disturbances in various tissues of habitual alcohol drinkers. In addition, alcoholic beverages supply calories, without supplying any other nutrients, thus "keeping you going" for the time being, without awareness of the serious, impending nutritional deficiency.

Calcium deficiency may blunt the outside/inside calcium difference in liver cells, thus augmenting the alcohol-induced liver cell damage and nutritional deficiency.

Once liver cells are injured, the first event is the entrance of calcium into the cell. Most likely the cell membrane is injured first, permitting calcium to enter the cell. Calcium activates protein splitting enzymes in the cell, leading to self-destruction. A calcium deficiency would exacerbate liver cell injury by increasing intracellular calcium. When enough calcium is taken by mouth, liver cell damage by alcohol might be minimized.

When we drink alcoholic beverages, blood vessels dilate bringing blood pressure down. This is followed by a rebound contraction of the blood vessel, so that blood pressure goes up. Habitual alcohol drinkers frequently have higher blood pressure than those who do not drink alcohol. By taking enough calcium, such a rise of blood pressure might be prevented to some extent.

Some people tolerate alcohol very well and never get drunk, but others are quite sensitive to alcohol. This is in part due to differences in the activity of *alcohol-detoxifying enzymes* in the liver. Heredity is one factor which determines the response and tolerance of an individual to alcohol. After drinking whiskey for many years, some people still have completely normal livers and enjoy healthy lives, whereas others suffer from liver diseases, even cirrhosis.

The field of immunogenetics has solved such an enigma. The thoroughness or meticulousness of immune surveillance is strongly influenced by heredity or disposition transmitted through generations. In some families, the immune surveillance system is too strict and occasionally indiscriminate, tracking down even normally functioning liver cells, only because of their strange behavior, like a nervous and irritated police officer.

When alcohol destroys a certain number of liver cells, the damaged liver cells are recognized as something unusual and unrelated to our own body. The normal immune system reacts reasonably by eliminating only these damaged liver cells, and the remaining liver cells stay intact. In a person with an overzealous immune system, the elimination process is extended to adjacent normal liver cells until the whole liver is damaged by autoimmunity—the immune system mistakenly directed against its own body.

Calcium deficiency may aggravate such a tendency through the activation of parathyroid hormone and blunting of the outside/inside calcium concentration gradient of lymphocytes. If the function of suppressor lymphocytes is thus compromised, the overzealous attitude of cytotoxic lymphocytes may be reasonably explained. Although the details of the complex interaction of the immune network are still far from clear, it is quite possible that a calcium deficiency might be responsible for the confusion of the immune network, aggravating alcohol-induced liver damage.

An small amount of alcohol, less than 30 grams a day, is said to be beneficial, decreasing the risk of cardiovascular disease and contributing to longevity. People drinking this amount of alcohol are said to suffer less frequently from fatal heart diseases than people drinking no alcohol at all. Combined with adequate calcium intake to make sure that the immune network is functioning properly, a reasonable amount of alcohol would not only be quite safe, but also be a useful tranquilizer.

(11) Smoking and calcium

Recently nothing good has been said about smoking. As research results emphasize the association between smoking and cancer, more and more people are giving up smoking. While all these risks of smoking really exist, nicotine contained in smoke also has a beneficial effect in that it stimulates the brain, possibly improving the memory. Without such a beneficial effect, smoking would probably not have become such a widespread practice. Substances other than nicotine, such as tar, are also responsible for the dangerous effects of smoking.

The brain is the center of the body's information network. Along the nerves running throughout the body like electric wires, messages are transmitted in all directions all the time. *Ganglia* are relay points for these messages, like telephone and telegraph stations. In order to facilitate and ensure the flow of information, substances called *neurotransmitters* are operating. Nicotine, the main component of tobacco, is one of these neurotransmitters. When nicotine enters the body as a result of smoking, a hormone called *vasopressin* or *antidiuretic hormone* is released from the brain to reduce the amount of urine. Noradrenaline is also released from the ganglia in response to nicotine, to raise blood pressure by contracting blood vessels. Smoking therefore excites the nervous system and raises blood pressure. Contraction of blood vessels interferes with blood circulation. The fingers and toes might become cold during smoking because of blood vessel contraction.

In *Buerger's disease*, arteries of the leg contract, decreasing blood flow to the leg and foot. While at rest, the leg manages with whatever blood is available. However, while walking over a long distance, a much larger amount of blood becomes necessary and the ordinary blood supply to the leg and foot becomes insufficient. Lack of oxygen causes muscle cramps and severe pain. On taking a rest, the need for blood and oxygen again decreases and pain subsides. This is called *intermittent claudication* of Buerger's disease. Too much smoking would cause blood vessel cramps and aggravate Buerger's disease. Calcium deficiency also causes blood vessel cramps. Since calcium deficiency and smoking may have similar effects, adequate calcium intake may somewhat decrease the damage caused by smoking and the aggravation of Buerger's disease.

A rise in blood pressure also occurs after smoking as well as in calcium deficiency. Adequate calcium intake may alleviate the rise of blood pressure through prevention of the entrance of calcium into blood vessels.

The most dreadful consequence of smoking is cancer. Cigarettes are notorious because the frequency of lung cancer—especially the *squamous epithelial variety*—is known to be proportional to the number of cigarettes consumed over years. Some forms of lung cancers, however, are apparently unrelated to smoking and occasionally even people who have never smoked develop lung cancer. Cancers of organs other than the lung, colon and stomach for instance, may also occur in response to smoking. The occurrence of cancer in places where the cigarette smoke does not contact the tissue directly may suggest the role of blood vessel contraction via nicotine or other agents. An adequate calcium intake may counteract the blood vessel contraction by preventing the entrance of calcium into the cell.

Smoking also decreases the appetite, inhibits the motility of the stomach and intestine, and increases gastric acid secretion, thereby aggravating gastric and duodenal ulcer. Calcium may antagonize the injurious effect of gastric acid. Calcium may also stimulate the secretion of calcitonin which has been found to be effective in decreasing gastric acid secretion and preventing cell injury with favorable effects on both gastric and duodenal ulcers.

5. Old Age and Calcium

(1) Dementia and calcium

Each of us becomes older, but the mode and speed of aging in various parts of the body are not the same. Muscle strength and skill of exercise decreases from around the age of twenty, and so swimming and gymnastics champions are usually young. Mental function such as judgment and adaptation, however, gradually mature in advancing age, along with experience. Salaried men retire at a certain age, but artists do not retire. Fine novels and paintings may be produced even after the age of eighty. Eventually, however, no one can beat the deterioration due to advancing age.

Mechanical memory reaches a peak during grammar school, but higher organized memory based on logical interrelationship and synthesis continues to improve. When you start to forget names and telephone numbers of people, you are sure you have grown up. Important people like you do not have to memorize small things. As long as you remember important things, you are well off.

Some people unfortunately keep forgetting things until daily life becomes difficult, causing trouble for other people. Senile dementia is said to represent the fourth most frequent cause of death in the United States. Along with heart disease, hypertension and stroke, senile dementia is one of the most dreaded diseases in the elderly people. The nature of senile dementia is poorly understood.

Artificial organs are recently being intensively studied. Artificial kidney and hemodialysis have given new life to people with renal failure. This is no doubt one of the most dramatic achievements of modern medicine. The artificial heart is also used. Among the many other organs of the body, the brain will perhaps be the only organ which cannot be replaced. When the brain of one person is replaced with the brain of another person, the individual is now a different person because all his past experiences, his memory and personality are lost. The computer is called the "electric brain" in China. Computers could certainly substitute for some of the brain's functions, but even the most complex computer on earth cannot entirely replace the complex functioning of the human brain. All the experiences in human life are stored in the brain, which therefore might be called life itself. That is why the decrease of brain function has an tremendous impact on daily life. The heart may be transplanted, but the brain will never be transplanted.

Memory represents only a small part of brain function. Emotion, will power and intellectual function make up a complex personality. Dementia is a deterioration of all these higher functions of the brain and is a serious menace to health.

Neither doctor nor drug cures the disease. Nature and the patient him- or herself bring about a cure. Will power to become well is indispensable to any form of medical treatment. In senile dementia, the loss of such will power makes recovery quite difficult. Patients with senile dementia are a heavy burden to both family and the nation.

What is the relationship between senile dementia and calcium? Attempts will be made to answer this question from the viewpoint of both environment and nutrition.

Diseases of the nervous system, brain, spinal cord and peripheral nerves are produced by many causes. A congenital anomaly is sometimes responsible, and environmental and nutritional factors are also important.

The Mariana volcanic zone extends from the Kii Peninsula in Central Japan via the Mariana Islands to New Guinea, across the Pacific. Strange diseases of the brain and spinal cord frequently occur along this zone. In Guam, a disease resembling a combination of dementia and Parkinson's disease is quite prevalent. In the Kii Peninsula of Wakayama Prefecture, central Japan, *amyotrophic lateral sclerosis* has occurred quite frequently. A strange disease called *Kuru* in New Guinea also gives rise to similar symptoms.

When Professors K. Kimura and Y. Yase of Wakayama Medical college visited the Kii Peninsula for an epidemiological survey of amyotrophic lateral sclerosis, they were surprised by the crystalline transparency of the water in the Koza River (Y. Yase et al., 1974). The water was so pure and transparent that no fish could live in it. The water contained practically no calcium. In the area of Guam with a high prevalence of *dementia-Parkinson disease complex*, the water was also exceptionally pure. As a matter of fact, the chemists at the National Institute of Health were upset because the water sample from Guam was indistinguishable from distilled water. In other parts of the world, natural water contains a moderate amount of calcium because the rain water takes up a large amount of calcium and other minerals while it runs through the soil. The unique property of volcanic ash along the Mariana volcanic zone is probably responsible for such a low calcium content in this water (Chen et al., 1984).

Although the calcium in drinking water accounts for only a small part of the calcium which comes into our body, if the calcium content of water is low, then the calcium content of the plants grown in that area will also be low. In addition, the meat of domestic animals which drink this water would also tend to be lower. Especially in older times when an area's food supply was narrowly limited, people living in these districts must have frequently been in the state of calcium deficiency. Calcium deficiency results in parathyroid overactivity and the release of calcium from bones. The calcium released from the bone tends to be accumulated in the brain and spinal cord, causing degeneration.

Amyotrophic lateral sclerosis is a dreadful disease in which the motor nerve system from the brain to the spinal cord degenerates. Arms and legs are paralyzed and muscles undergo atrophy. Finally, respiratory muscles in the chest wall become paralyzed and the patient dies. When the brains and spinal cords of decreased patients were examined, very high calcium contents were detected. Other metals such as aluminum and manganese were also increased. Calcium was probably the first to be deposited, as a result of the Calcium Paradox in calcium deficiency, causing degeneration of the brain and spinal cord. Other metals would then start to accumulate to cause a vicious cycle and a further increase of calcium deposition. When experimental rats were maintained on a diet containing little calcium, they became paralyzed and weak, just like the patients with amyotrophic lateral sclerosis, with degeneration of the brain and spinal cord because of high calcium deposition there. Calcium deficiency was thus apparently responsible for these dreadful diseases of the brain and spinal cord.

The function of the kidney decreases with advancing age, and synthesis of the active form of vitamin D also falls. The degree of decrease in renal function is mild, seldom requiring dialysis, but is nevertheless steadily progressive. Consequently, gut calcium absorption falls. Calcium deficiency therefore becomes more and more pronounced as one grows older. Parathyroid hormone must therefore be used more frequently to take calcium out of the bone. Indeed, serum parathyroid hormone increases with age. Calcium then accumulates in the brain as well as in the blood vessels. Patients with renal failure have a physiology similar to elderly people. Calcium was found to be increased in the brains of patients who died of senile dementia, Alzheimer's type, as well as patients with amyotrophic lateral sclerosis, and those subjected to long term hemodialysis.

Cerebral arteriosclerosis and Alzheimer's disease are two important causes of senile dementia. Both may be caused by calcium deficiency, release of calcium from bone, and an increase of calcium in the soft tissue—brain and the blood vessels. Patients with Alzheimer type dementia were found to have taken less milk when they were young, compared to those with other diseases, according to a report from the Yokufūkai Home for the aged in Tokyo.

(2) Cataract and calcium

Cataract is a cloudiness of the eye lens caused by a white precipitate. An increase of calcium in the lens content is also usually found. With advancing age, more and more people suffer from cataracts. Unless the affected lens is removed by surgery and eyeglasses are used, the vision is progressively impaired until it is completely lost. The reason calcium enters the lens is not fully understood. People with diabetes mellitus frequently suffer from cataract. It is quite possible that the principle of Calcium Paradox again operates here. Calcium deficiency in elderly people might be causing a flood of calcium throughout the body, including the eye lens.

(3) Longevity and calcium

Human life is a struggle against calcium deficiency, involving everyone, from the baby in the crib to the elderly. The aging process is inseparable from calcium deficiency. Compared to fish in water, we are always more deficient in calcium. As we grow and become older, the degree of calcium deficiency becomes even more pronounced.

As we become deficient in calcium, more and more parathyroid hormone is secreted, to release calcium from the bone. Calcium from the bone enters soft tissues such as the blood vessel wall and brain. Parathyroid hormone also changes the property of the membrane of many cells to increase the entrance of calcium into the cell. Since maintenance of the vast difference in calcium concentration between outside and inside the cell up to 10,000 is necessary for the maintenance of adequate cell function, parathyroid hormone overactivity is not only deleterious to the bone, which loses calcium, and blood vessels and brain, where calcium accumulates, but also interferes with most of the cells in the human body. Calcium shift takes place in the *macroscopic* field from the bone to soft tissues and also in the microscopic field from outside to inside of the cell, by the action of parathyroid hormone in response to calcium deficiency. Adequate nutritional calcium intake would be capable of minimizing or preventing all these effects. Calcium stays in the bone to keep it strong and to prevent excessive calcium from accumulating in the soft tissues. Once calcium is deposited in the tissues, it is difficult, if not impossible, to remove. One ounce of prevention is worth a pound of treatment. Calcium release from the bone and deposition in soft tissues should be prevented before it occurs. Calcitonin and the active form of vitamin D represent therapeutic possibilities to prevent such a calcium shift, in addition to taking enough calcium as food and calcium supplements.

One of the causes of progressive calcium deficiency in aging is the decline of kidney function to produce the active form of vitamin D. In addition to the decrease of active vitamin D to help gut calcium absorption, the gut itself may lose its ability to absorb calcium and other nutrients.

In order to evaluate the importance of the kidney in longevity, Dr. D. N. Kalu, of the University of Texas, maintained rats on a diet with low protein and usual calcium content, keeping calcium intake adequate in both groups (Kalu et al., 1984). Rats consuming a standard protein intake died sooner than those with restricted protein intake. Before they died, parathyroid hormone rose sharply, indicating a terminal renal failure causing deficient production of the active form of vitamin D and a decreased gut calcium absorption, stimulating parathyroid hormone secretion. Rats maintained on the low protein diets never developed such a rise of parathyroid hormone in the blood, suggesting that their kidneys continued to function well. A protein load is thus unfavorable for the kidney, and avoiding excessive protein may contribute to longevity, through preservation of kidney function and the ability to produce active vitamin D. Vitamin D-stimulated favorable calcium absorption in the intestine is therefore an important key to longevity through preserving the delicate balance of calcium between bone and soft tissue, and between outside and inside of the cell.

(4) Calcium paradox of aging

The behavior of calcium in our bodies is a paradox. The less calcium we ingest in food, the more calcium is removed from the bone, via an increased stimulation for parathyroid hormone secretion. Over fifty years ago, when we did not know about primary hyperparathyroidism, we had no treatment for patients with this disease. Left untreated, these patients suffered from the tragic development of *osteitis fibrosa cystica*, with a loss of calcium from the bone, and consequently the deformity of the whole skeletal system. The bone would become soft enough to be easily cut with a knife. The calcium removed from the bone by parathyroid hormone was deposited in the soft tissue—blood vessels, kidneys and brain.

This process occurs to a milder degree when insufficient amounts of calcium are ingested. Thus, the less calcium we take in, the more the calcium gets deposited in the blood vessels. This is called the Calcium Paradox of Aging.

When we ingest more calcium, the parathyroid gland stays "asleep," and the bone calcium remains secure. No calcium gets deposited in the blood vessels. Therefore the more calcium we take, the less the calcium gets deposited in blood vessels.

What evidence do we have for this strange, paradoxical behavior of calcium?

Chronic renal failure presents a good example of this calcium paradox. In renal failure, the kidney is unable to make the active form of vitamin D. Calcium absorption in the gut markedly falls. The parathyroid gland responds by secreting an excessive amount of parathyroid hormone, creating a condition known as secondary hyperparathyroidism. Bone calcium will be withdrawn and the calcium in blood vessels, nerves and the brain will be increased. Studies of rats with laboratory-induced renal failure showed that the amount of *labelled Ca*^{45} was tremendously increased in the aorta, heart and kidney, while the amount of labelled Ca^{45} in the bone decreased.

Fig. 25 Rats given sulfacetyl thiazole (SAT), a poorly soluble sulfonamide which causes kidney damage, showed accumulation of isotopic calcium in the aorta, heart and kidney, but less calcium accumulation in the bone. In parathyroidectomized (PTX) animals, the same dose of SAT caused no such accumulation, indicating a calcium shift from bone to the cardiovascular system through the action of parathyroid hormone.

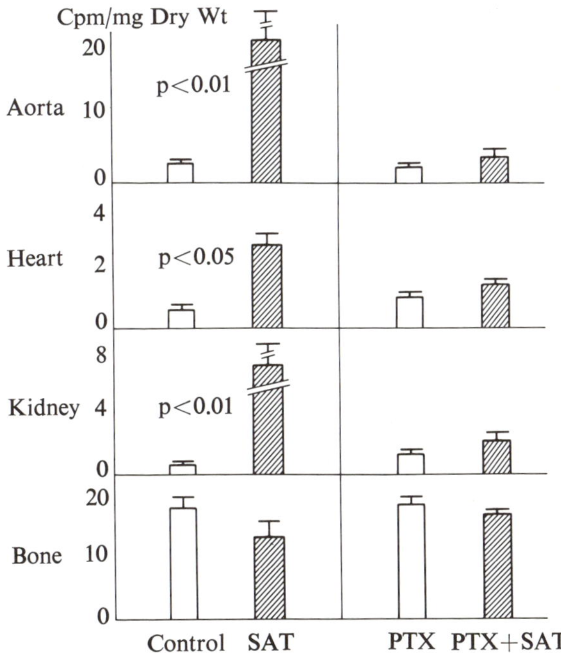

However, in another group of rats with renal failure, after parathyroidectomy (PTX), no such increase of Ca^{45} was found (K. Okano et al., 1970).

Clinically, aortic calcification is more frequently observed in osteoporotics than in non-osteoporotics (Anderson et al., 1964). This observation led Dr. A. Elkeles, a British radiologist, to propose the *calcium shift theory* describing the blood vessel, with advancing age, as kidney function decreases (Erkeles, 1957).

In an epidemiological survey of two regions in the Kii Peninsula of Central Japan, the incidence and degree of spinal compression fracture paralled that of aortic calcification. In addition, in the mountain area, which had less sunshine exposure and a lower calcium intake, the people were shorter, complained more of lumbago, and had lower serum phosphorus and cholesterol and thinner bones than the people in the seacoast district (Fujita et al., 1977, 1984).

As might be expected, the incidence of amyotrophic lateral sclerosis was extremely high in the mountain district of the Kii Peninsula. Nerve degeneration was the result of chronic low calcium intake, with deposition of calcium in the brain and spinal cord.

Fig. 26 Aging and calcium metabolism.
The aging kidney produces smaller amounts of active vitamin D. An aging gut absorbs less calcium. The resulting parathyroid overactivity takes calcium away from the bone. This calcium is deposited in the blood vessel and brain.

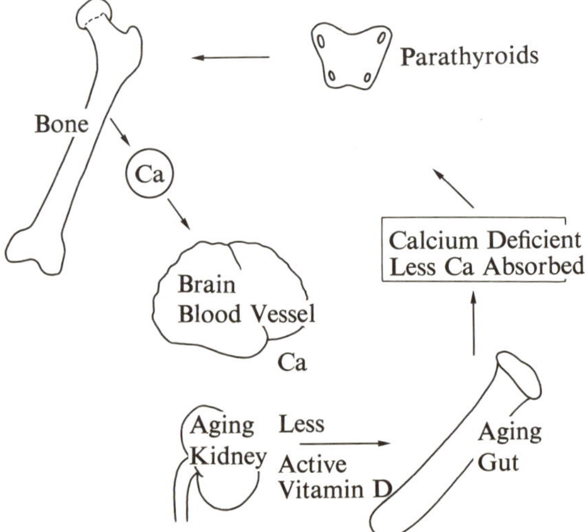

This principle of the Calcium Paradox of Aging from calcium deficiency and hyperparathyroidism thus appears to be widely applicable to understanding the pathogenesis of diseases associated with old age. Calcium deficiency, transient *hypocalcemia*, and secondary *hyperparathyroidism* can all be caused by both lack of dietary calcium intake and reduced activity of the aging intestine and kidney.

As the individual ages, absorptive capacity is reduced, and intestinal

Fig. 27 Calcium paradox of aging.
When too little calcium is ingested, and vitamin D and ca lcitonin are deficient, more calcium comes out of the bone and is deposited in the blood vessel. When sufficient calcium is taken by mouth, on the contrary, or when vitamin D and calcitonin are acting satisfactorily, less calcium comes out of the bone and scarcely any unnecessary calcium is deposited in the blood vessel.

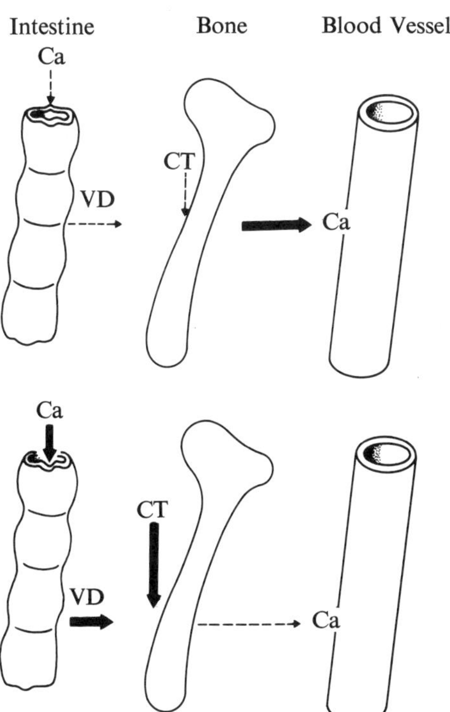

calcium absorption is decreased; kidney function also decreases, with a corresponding reduction in active vitamin D production, and further decrease in calcium absorption. This will at first cause a loss of calcium from the bone and osteopenia. The calcium released from bone enters blood vessels, contributing to the development of hypertension and arteriosclerosis, or is deposited in the brain, contributing to the onset of senile dementia. Parathyroid hormone may also transfer calcium from the outside to the inside of the cell (Borle, 1972). This increase of intracellular calcium may stimulate cell proliferation, and eventually, malignant cell growth. Blunting of the outside/inside

Fig. 28 Deformity of the spine (upper panel) and aortic calcification by age in males (M) and females (F) in mountain (black bar) and seacoast (open bar) districts. Close parallelism between spinal deformity and aortic calcification is evident.

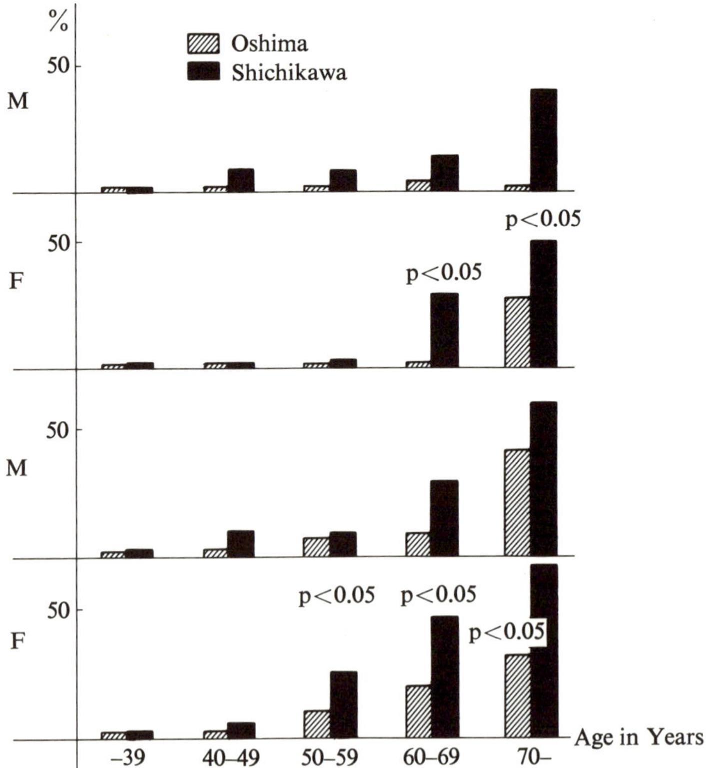

ratio of cell calcium may explain the general decrease of cellular activity in aging. A decrease in the activity of beta cells of the pancreas may help to explain deficient insulin secretion and/or diabetes mellitus. Decrease of the activity of the immune cells may explain elderly people's increased susceptibility to infection.

Longevity is what everybody wishes but not everybody can achieve. There are so many things which influence the life span. Heredity is an important factor. By avoiding all diseases of old age—hypertension, arteriosclerosis, çerebrovascular disease, cancer, pneumonia, etc.—our life expectancy could no doubt be extended.

At the Katsuragi Geriatric Hospital in Osaka, Japan, all the patients are above the age of 65. Here a retrospective survey was conducted on the relative contribution of various clinical variables to the outcome of hospitalization. Strangely enough, age, duration and hospitalization, serum total protein, serum alfremin, GOT, GPT, serum cholesterol, serum uric acid and serum sodium had very little

Fig. 29 Diseases of old age and calcium metabolism. Decreased calcium intake and absorption cause secondary hyperparathyroidism and osteoporosis. Shift of calcium from bone to blood vessels causes hypertension and arteriosclerosis, and calcium deposition in the brain may induce senile dementia. Malignancy, immune cell hypofunction and diabetes may be due to blunted inside/outside concentration differences of calcium in the cell.

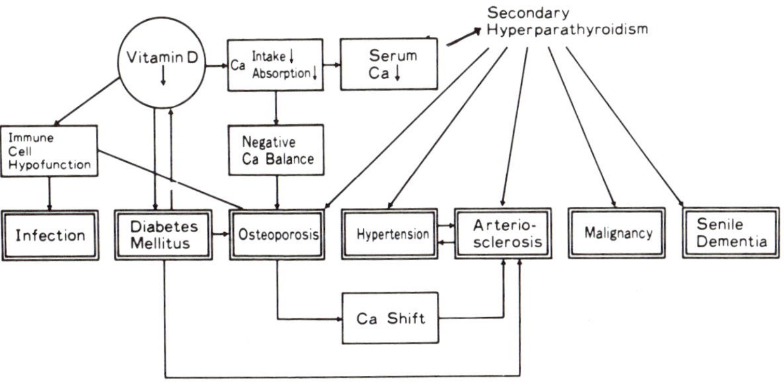

influence on the outcome of the patient. Higher levels of serum creatinine were associated more frequently with fatal outcomes at the hospital.

Fig. 30 Laboratory tests and outcomes of hospitalization (Mean±SD).

	Improved (N=158)	Died (N=432)	Difference
Serum Creatinine	1.0±0.2	1.2±0.7	p<0.01
Serum Uric Acid	4.7±1.7	5.2±2.5	p<0.05
Serum Ca	9.2±0.5	8.9±0.7	p<0.001
Serum Ca Corrected by Albumin	9.7±0.4	9.7±0.5	NS
Serum Total Protein g/dl	6.5±0.5	6.3±0.7	p<0.001
Serum Albumin g/dl	3.4±0.4	3.1±0.5	p<0.001
GOT U	29±34	34±46	NS
GPT U	18±32	20±33	NS
Cholesterol	189±50	180±45	p<0.05
Na mEq/L	140±3	138±4	p<0.001

(mg/dl unless specified)

Most impressive was the contribution of serum calcium. Higher serum calcium levels were associated with a better chance of patient survival and hospital discharge. This study included more than 500 patients. Although the causes of death are quite variable and many factors are involved, this study indicated that a higher serum calcium level (within the normal limits) may be protective against the dangers of hypocalcemia and accompanying secondary hyperparathyroidism (Fujita, 1984).

Fig. 31 **Discriminant function between improvement and fatal outcome. Positive values (right) contribute to improvement and negative values (left) to death. The higher the serum creatinine, or the poorer the kidney function, the poorer the chance for survival. The higher the serum calcium, on the contrary, the greater the chance for improvement.**

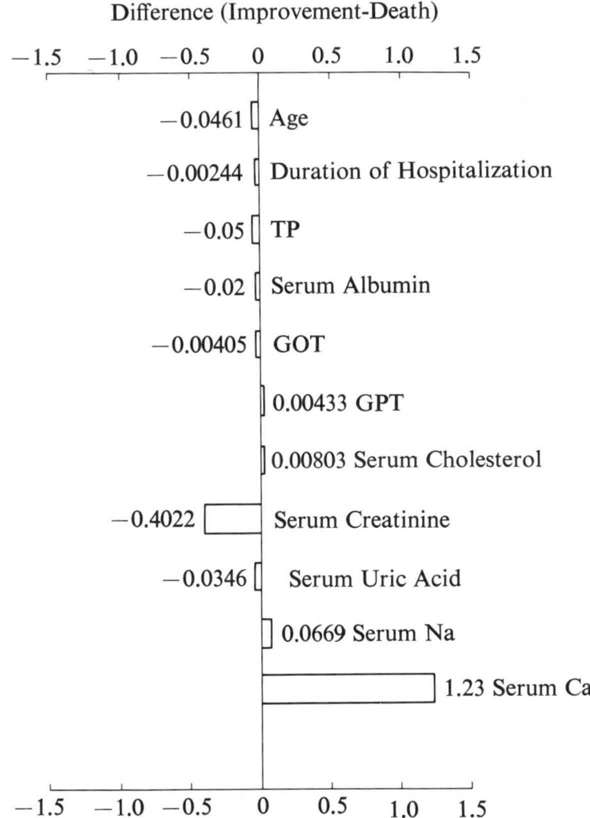

3. How to Solve the Problem of Calcium Deficiency

1. How Is Calcium Absorbed?

Calcium is absorbed by two different mechanisms. One is simple diffusion or passive absorption, and the other is active absorption. Simple diffusion always occurs from regions of higher to those of lower concentration, in an attempt to equilibrate the distribution of calcium between two compartments. This occurs spontaneously or without spending energy, like water streaming down a mountain. Active transport, on the contrary, is like bringing water up to a mountain top using a car or lift. Since this process is going against gravity, some energy such as gasoline, electricity or manpower is required. When no energy is spent, no water goes up. Active transport is therefore easily controllable. With more effort, more work is done.

In the human intestine, active transport of calcium mainly occurs in the duodenum and upper *jejunum*, the beginning part of the small intestine. Vitamin D mainly acts on this part of the intestine to increase the synthesis of calcium-binding protein. Whenever vitamin D acts on the intestine to stimulate calcium absorption, calcium-binding protein is increased. Formerly, calcium-binding protein was thought to carry calcium across the cell, from the mucosal to serosal side, but the situation is apparently not so simple. Calcium is absorbed in response to vitamin D in a matter of minutes, much faster than the synthesis of calcium-binding protein. Since an increase of intracellular calcium is toxic to any cell, the calcium-binding protein may alleviate the sudden increase in intracellular calcium by binding with calcium entering the cell.

In other parts of the intestine, vitamin D may also work, but not as dramatically as in the duodenum. When dietary calcium intake has been low for many years, the intestine adapts itself to calcium deficiency by an increased efficiency of calcium absorption. Active vitamin D synthesis is increased in response to calcium deficiency, so that the calcium absorption from the duodenum and upper jejunum is naturally increased. The lower part of the intestine, ordinarily engaged in calcium absorption only by passive diffusion, also adapts itself to calcium deficiency by increasing calcium absorption. The contribution of this part of the intestine might not be as great as that of the upper part of the intestine, but is nevertheless quite important to the total calcium absorption process.

2. How Is Intestinal Calcium Absorption Controlled?

Vitamin D is the single most important factor for controlling calcium absorption. As stated previously, it is the final active form of vitamin D, 1,25$(OH)_2$ vitamin D, that stimulates intestinal calcium absorption directly. Some of the vitamin D used by our body is already present in food as vitamin D, but most of it enters the body as previtamin D. Previtamin D has to be converted in the skin to become vitamin D. The first step in the activation of vitamin D takes place in the liver, and produces 25(OH) vitamin D. Finally, in the kidney, 1,25$(OH)_2$ vitamin D is synthesized. This is the final form of active vitamin D and this step is subject to various metabolic controls. The most important controlling factor for 1,25$(OH)_2$ vitamin D synthesis in the kidney is parathyroid hormone. Whenever necessary, parathyroid hormone stimulates the kidney to make more active vitamin D. When either serum calcium or serum 1,25$(OH)_2$ vitamin D levels are low, parathyroid hormone secretion is stimulated. When dietary calcium intake is low for a long period, more and more parathyroid hormone is secreted. This is due in part to repeated minor degrees of decreased serum calcium levels. Through the action of parathyroid hormone stimulating the kidney to produce more active vitamin D, a low calcium intake increases active vitamin D production. Even when serum calcium levels are normal, the secretion of parathyroid hormone is stimulated when the amount of active vitamin D in blood is low. An increased stimulation of parathyroid hormone will thus increase the production of active vitamin D.

When dietary calcium intake is sufficient, and serum calcium is raised, or when the amount of active vitamin D in the blood is high, parathyroid hormone secretion decreases and less active vitamin D is produced.

Production of the active form of vitamin D is also controlled by serum phosphorus levels, independent of parathyroid hormone and calcium levels. High blood phosphorus levels directly inhibit the kidneys' production of 1,25$(OH)_2$ vitamin D, and thereby reduce calcium absorption. Since blood phosphorus is not as precisely controlled as blood calcium, as more phosphorus is ingested, serum phosphorus rises. Phosphorus may also decrease calcium absorption directly by precipitating calcium as an unabsorbable salt.

Calcitonin may sometimes stimulate 1,25$(OH)_2$ vitamin D production by the kidney to indirectly increase intestinal calcium absorption but it also tends to directly decrease intestinal calcium absorption by inhibiting intestinal motility. Thyroid hormone also decreases intestinal calcium absorption. Adrenal glucocorticoid reduces calcium absorption, possibly through interfering with the action of vitamin D. Among the food components, lactose facilitates calcium absorption. *Lactose intolerance* is a hereditary disorder in which an enzyme (lactase) capable of splitting lactose into glucose and galactose is lacking. As a result, lactose-containing food such as milk and other dairy products cannot be readily digested. This may lead to diarrhea or abdominal pain whenever such food is taken. Since lactose facilitates calcium absorption, lactose intolerance is a handicap for calcium absorption and availability. In order to facilitate calcium absorption in subjects with lactose intolerance, lactase as a drug may be used as a supplement for the lacking enzyme. Even if lactose intolerance is present, this is not always serious, because diarrhea does not necessarily occur unless a large amount of milk is ingested.

Contrary to the common notion, glucose does not inhibit, but rather may facilitate calcium absorption. Dietary fibers and some organic acids such as *phytic acid* tend to reduce calcium absorption. Even if the calcium content of certain vegetables may appear high, coexistence of oxalate and phytate can interfere with absorption and utilization. Adaptation of the intestine to various levels of dietary calcium intake is one of the wonders of the human body. Utilizing many of the control mechanisms described above, the gut absorbs more calcium when dietary calcium intake is low, and much less calcium when abundant dietary calcium is taken. Children and pregnant women absorb calcium very efficiently. This is important because they need a lot of calcium. As one gets older, less and less calcium is absorbed, because the intestine itself loses its efficiency; in addition, less active vitamin D is available. Estrogen facilitates calcium absorption, possibly through the stimulation of active calcium synthesis by the kidney. By the help of estrogen, calcium in the food is absorbed and utilized more efficiently than in its absence. One of the reasons why the calcium requirement increases after menopause is the dramatic decline in estrogen.

3. Calcium Balance and Requirement

When a certain amount of calcium is ingested, part of it, up to 70 percent in babies and 30 to 40 percent in adults, is absorbed by the intestine and the remaining part passes through the intestine to be excreted in the feces. A portion of the absorbed calcium is excreted in the urine, but the rest remains in the body at least for a time. Without considering the movement of calcium within the human body except for calcium entry and excretion, we define the total calcium intake to consist of all ingested calcium, and the total calcium output to consist mainly of urinary and fecal calcium excretion.

When the total calcium intake is larger than the total calcium output, our body is said to be in positive calcium balance, gaining and storing some calcium. On the contrary, when total calcium output exceeds the total calcium intake, we are in negative calcium balance, losing some calcium every day.

The recommended dietary allowance (RDA) for calcium is different from the calcium requirement. In every country, the nutritionists work hard to determine the RDA, because such a figure has a profound influence on the public. The RDA for a nutrient is the level of intake at which most of the people would be safe from a nutritional deficiency as well as from an excessive intake. In order to establish the RDA, many nutritional balance studies are done on subjects of varying ages, both males and females. A margin of safety is also added to the minimum requirement. Adequate levels of calcium intake naturally vary according to age, sex, modes of living and degree of exercise. These factors influence the availability of calcium and therefore the adequate level of calcium intake. Each nation has set its own RDA for calcium. Many nations have also set different levels of RDA for various age and sex groups.

Since women are more susceptible to the dangers of a negative calcium balance because they usually eat less than men and they spend calcium for childbearing and lactation, they often have more difficulties in maintaining adequate calcium intakes. Strict caloric restriction inevitably lowers the dietary calcium intake, unless mineral supplements are taken. In the United States, the average dietary calcium intake is lower than the recommended dietary allowance (RDA), especially for women. Elderly women are taking even less calcium, perhaps as little as 60 percent of the RDA. Since elderly people absorb less calcium from the intestine, low calcium intake would be even more deleterious for them. Elderly people therefore are at a much greater risk of calcium deficiency than younger age groups. In spite of the efforts of nutritionists and the government to set a reliable RDA, many people either do not understand the importance of RDA or somehow fail to follow the recommendations.

Fig. 32 Recommended dietary allowances of calcium (mg) in various countries.

Country	Male	Female	Pregnant Women
U.S.A.	800	800	1,200
Argentina	700	600	2,100
Britain	500	500	1,200
Italy	500	500	1,200
India	450	450	1,000
Canada	800	700	1,250
Sweden, Norway	600	600	1,000
Soviet Union	800	800	1,500
Korea	600	600	1,000
Japan	600	600	1,200
Taiwan	600	600	1,100
Czechoslovakia	800	800	1,500
China	600	600	1,500
France	800	800	1,000
FAO	450	450	1,100

Urinary calcium excretion is also a factor in determining calcium balance. When the protein content of the diet is too high, urinary calcium excretion may be increased and the calcium balance might become more negative. Phosphorus, possibly by reducing calcium absorption, decreases urinary calcium excretion. Thiazide, a drug which increases urinary sodium excretion and reduce blood pressure, also decreases urinary calcium excretion, and has been used to prevent kidney stones.

Even if the calcium balance is somehow maintained, the calcium intake is not necessarily adequate. Let us consider our income. We could live on $500 a month if we have to, by cutting down on expenses. This does not mean that $500 a month is an adequate income. With $5,000 a month, we could also maintain a balance, at a much higher level than with an income of $500. Similarly, we could get along much better with a calcium intake higher than the recommended daily allowance. The fact that we could somehow live on a daily calcium intake of 800 mg is by no means a proof that this is the adequate calcium intake.

Pregnant and lactating women, as well as growing children, have increased requirements for calcium. As one grows older, one should take more calcium because the intestinal absorption of calcium decreases and the risk of osteoporosis constantly increases, more in females, but certainly in males also. We grow up and become older, so that there is no moment in life when we do not need calcium. One gram of calcium a day is a safe rule of thumb, with additions if necessary for growth, childbearing, and diseases.

Why, then, is the RDA necessary? It is always nice if we have a practical target not too far away, apart from the ultimate goal. Although many of us are not even meeting the RDA, it certainly appears to be possible to meet the RDA.

There are controversies on the RDA from opposite directions. Some nutritionists mindful of calcium deficiency feel that the RDA for calcium is still too low, whereas others who are very much afraid of excess think that even the current RDA is too high. Since the RDA apparently represents the consensus of the opinions of many scientists, we are well advised to try to meet this level of intake. It is up to the individual, however, to interpret the RDA in the light of new scientific data and adjust his or her intake accordingly.

4. What Kind of Calcium Containing Foods Are Most Desirable?

Food is naturally the most convenient and useful source of any nutrient, including calcium. Calcium-rich food may be classified into four major categories; dairy products, fish and shellfish, seaweeds and vegetables. Intestinal absorption and availability of calcium varies widely among these groups. Among our daily foods, milk and other dairy products contain the largest amount of calcium. Milk contains as much as 240 mg calcium per 200 cc or 1 glass. If you drink 5 glasses of milk, that will already provide 1.2 g calcium. One cup of

Fig. 33 Foods rich in calcium.

Kind of Food	Name of Food	Ca (mg) in 100 g edible portion
Dairy Products and Eggs	Cow's milk	118 (whole) 121 (skim)
	Yogurt	111 (whole) 120 (skim)
	Ice Cream	146
	Cheddar Cheese	761
	Swiss Cheese	936
	Cottage Cheese	96
	Blue Cheese	315
	American Cheese	697
	Egg	47
Fish and Shellfish	Canned Sardine	435
	Eel	95
	Shrimp, raw	120
	Sweetfish (Ayu)	270
	Oysters, raw	94
	Clams, raw	69
	Mackerel, canned	260
	Salmon, red, canned with bones	259
	Kale, raw	249
Seaweeds	*Wakame* (Seaweed) (dry)	960
	Hijiki (dry)	1,400
	Tangle (*Konbu*) (dry)	710
	Nori (*Yakinori*) (dry)	410
Soy Products	Soy Milk	31
	Tofu (Soybean cake)	128
	Natto (Fermented soybeans)	90
Seeds	Sesame	1,200
Vegetables	Broccoli, raw	103
	Spinach, raw	93
	Turnip greens, raw	246
	Collard greens, raw	183
	Mustard greens, raw	203
	Cabbage	49
Others	Almond	234
	Blackstrap Molasses	685
	Orange	30

Values Derived From: U.S. Dept. of Agriculture, Agricultural Research Service, *Nutritive Value of American Foods*, Agriculture Handbook No. 456, Washington, D.C., 1975
Kagawa, Aya, editor, *Standard Tables of Food Composition in Japan*, Woman's College of Nutrition, Tokyo, 1986

yogurt (200 g) and one cup of ice cream (150 g) contain about 240 mg calcium. Similar amounts of calcium may be supplied by 40 g of cheese. There is, however, a limitation in increasing the calcium intake by taking more and more dairy products, because phosphorus also contained in these products may interfere with calcium availability. Whole fish with bones, like canned sardines, usually contain abundant calcium and phosphorus although the Ca: P ratio varies among various fishes. Seaweeds, green vegetables, sesame seeds, almonds and soybean products such as tofu contain abundant calcium. Vegetable sources have the advantage of a lower phosphorus content, but fibers and phytates may interfere with calcium absorption. Mother's milk contains less calcium than cow's milk, but calcium in mother's milk is more efficiently absorbed than the calcium in cow's milk. Lactose, the sugar contained in milk, facilitates calcium absorption.

Fig. 34 Rate of calcium absorption.

Food	% Absorbed
Milk	53
Calcium Carbonate	42
Small Fish with Bones	34
Vegetables	18

For selection of food as a calcium source, not only the calcium content, but also other components should be examined. Since protein increases urinary calcium excretion, foods containing too much protein should be avoided, although the simultaneously high intake of protein and phosphorus tends to make the negative influence on calcium balance milder. Sugar does not remarkably influence urinary calcium excretion. Fiber may be recommended for constipation, diabetes mellitus and overweight, but fiber also interferes with calcium absorption and should not be taken in excess. An adequate calcium/phosphorus ratio is important, and too much phosphorus should be avoided because phosphorus interferes with calcium absorption.

Calcium may be supplied from bone powder, egg shells and oyster shells. Bone powders contain too much phosphorus to insure adequate calcium availability.

Among various calcium supplements, the proportion of calcium is quite different. Simple inorganic compounds such as calcium carbonate and calcium oxide have high calcium contents and are usually highly absorbable.

Fig. 35 Calcium: Phosphorus ratios and calcium contents of selected foods (Per 100 g edible portion).

Calcium-rich Foods (Ca: P=1: 2 and below)			Calcium-phosphorus Balanced Foods (Ca: P=1: 2.1–1: 7)			Phosphorus-rich Foods (Ca: P=1.7 and over)		
	Ca: P	Ca (mg)		Ca: P	Ca (mg)		Ca: P	Ca (mg)
Bread, cracked wheat	1: 1.5	88	Italian bread	1: 4.6	14	Buckwheat, noodle,	1: 8.9	9
French	1: 2.0	43	Macaroni, cooked	1: 6.1	17	boiled		
Raisin	1: 1.2	72	Rice, brown,					
Rye	1: 1.9	76	cooked	1: 6.2	12	Corn, cooked	1: 29.4	5
White	1: 1.1	86	Rice, white cooked	1: 2.7	8			
Broccoli	1: 0.7	88				Potato	1: 7.5	5
Cabbage	1: 0.6	49				Mushroom	1: 20.3	6
Carrot	1: 1.0	33	Beer	1: 6.0	5			
Collard greens	1: 0.3	152	Chestnuts, boiled	1: 3.0	22			
Cucumber	1: 1.1	24				Bonito	1: 27	10
Kale	1: 0.3	187	Soybeans, fermented,			Cuttlefish	1: 9.4	18
Lettuce	1: 1.1	20	*Natto*	1: 2.1	90	Mackerel	1: 7.3	22
Mustard greens	1: 0.2	137	Soybeans, cooked	1: 2.5	73	Sea bream	1: 10	15
Onion	1: 1.3	27				Tuna, canned in		
Parsley	1: 0.3	203				water	1: 11.8	16
Radish, long white	1: 0.7	30						
Spinach	1: 0.4	93	Bamboo shoots	1: 2.8	18			
Sweet potato	1: 1.5	23	Bean sprouts	1: 3.4	19	Bacon, cooked	1: 18	13
Tomato	1: 2.0	12	Garlic	1: 6.0	33	Beef	1: 16.2	8
Turnip greens	1: 0.2	100	Lotus root	1: 3.3	18	Chicken	1: 22.4	9
Welsh onion	1: 0.4	47	Pepper	1: 2.3	7	Ham, Pork	1: 19.6	8
Hijiki seaweed	1: 0.07	1,400						
Nori, toasted	1: 1.5	410	Banana	1: 3.1	6			
			Peach	1: 2.1	8			
Konnyaku, Devil's			Crab	1: 2.8	60			
Tongue	1: 0.1	43	Fish paste, *Kamaboko*	1: 2.4	25			
Taro	1: 1.9	22	boiled					
Tofu, soybean curd	1: 1	128						
Butter	1: 0.8	20	Mackerel pike	1: 2.1	75			
Cheese, cheddar	1: 0.6	750						
Ice cream	1: 0.8	146	Lobster	1: 3.0	65			
Milk	1: 0.8	118	Egg	1: 3.8	48			
Yogurt	1: 0.8	120						
Mayonnaise	1: 0.1	18						
Blacktrap molasses	1: 0.1	685						
Pickled plum	1: 0.9	24						
Apple	1: 1.6	6						
Lemon	1: 0.6	17						
Tangerine	1: 0.4	30						
Herring, canned	1: 2.0	147						
Salmon, red, canned								
with bones	1: 1.3	259						
Sardine, canned	1: 1.1	437						

These compounds, however, may be irritating to the stomach. Calcium phosphate may have a disadvantage of phosphate interfering with the absorption of calcium. Organic compounds such as calcium lactate and gluconate have lower calcium content, necessitating ingestion of larger amounts.

Calcium carbonate, the most widely used calcium supplement, appears to be the best supplement currently available. Calcium may be supplied from bone powder, egg shells and oyster shells. Bone powders contain too much phosphorus to insure adequate calcium availability. Oyster shell, mainly consisting of calcium carbonate, is fine, but a preparation better than simple oyster shell powder is available. Such an "active, absorbable form of calcium" has a very high electric conductivity—10,000 μs/cm compared to 80 μs/cm in simple oyster shell powder.

Fig. 36 Proportion of calcium in calcium compounds.

Compounds	Ca Content (%)	Grams of Compound Necessary to Provide 1 gm Ca
Calcium Carbonate	45	2
Calcium Oxide (Active Absorbable Calcium)	50	2
Calcium Phosphate	32	3
Bone Powder	32	3
Calcium Lactate	13	7
Calcium Gluconate	10	10
Calcium Aspartate	9	10

This specially made, active form of calcium from oyster shell powder is better absorbed than calcium carbonate, and it has been tested in patients with low parathyroid hormone activity following surgery for primary hyperparathyroidism. When parathyroid hormone is deficient, the kidney makes less active vitamin D, and the intestinal absorption of calcium decreases. The active form of calcium may be absorbed better than calcium carbonate, despite low vitamin D activity. Active absorbable calcium has been shown to be better absorbed than calcium carbonate in vitamin D-deficient rats. Thus, active absorbable calcium appears to be utilized much better than an ordinary calcium preparation such as simple powdered oyster shell or calcium carbonate. Active absorbable calcium assumes a characteristic crystalline shape, unlike calcium oxide.

Active absorbable calcium not only raises serum calcium more effectively than calcium carbonate, but also suppresses parathyroid hormone levels in blood much better. Even in patients with renal failure and low serum active vitamin D levels, active absorbable calcium is absorbed through the gut to normalize serum calcium and decrease parathyroid hormone in blood to the normal level. Active absorbable calcium thus takes away the Calcium Paradox and restores the deranged calcium metabolism to normal whenever a calcium deficiency is present.

5. How Much Calcium Should We Take?

Japanese people consume approximately 600 mg of calcium a day and, on average, Americans consume slightly more. There are, however, many factors influencing the availability of calcium such as the dietary intake of phosphate, protein and vitamin D, exposure to sunshine, degree of exercise, and so on. Thus, it is difficult to determine an average satisfactory intake. It would be wise to recommend a little higher amount just to be on the safe side. In each nation, children and younger subjects eat more calcium than elderly people. Young people may be taking an amount of calcium corresponding to the RDA, but many elderly people fail to do so.

Is it dangerous to take too much calcium? Everything has a limitation and calcium is no exception. There is no reason to believe that a limitless calcium intake is safe. However, calcium is a very special substance for our body and its metabolism is carefully controlled. If we compare it to another nutrient, we can easily understand. For example, taking too much potassium raises the serum potassium level and a potassium deficiency lowers serum potassium. This is because the intestinal absorption of potassium largely depends on passive diffusion. Calcium is different. Intestinal calcium absorption is carefully controlled and adjusted. When calcium intake is insufficient, the intestine absorbs more calcium by both active and passive processes, depending on vitamin D and other factors. On the contrary, when too much calcium is ingested, calcium absorption from the intestine is markedly decreased. Through such careful control by the intestine, the serum calcium level scarcely rises, even after ingestion of a large amount of calcium. When we regularly consume a large amount of dietary calcium, the intestine adjusts itself accordingly. However, there is always a possibility that an excess intake of one nutrient may interfere with absorption of others. Calcium may interfere with the absorption of cholesterol and other lipids, among others. While this appears to be favorable in terms of preventing excess fat, no one knows whether or not absorption of other important substances may also be inhibited by an excessive calcium intake.

Prior to the advent of modern pharmacology, gastric and duodenal ulcers were treated with alkali and huge amounts of milk to counteract gastric acid secretion. Patients sometimes developed *milk-alkali syndrome* or a rise of serum calcium. In this unusual situation in patients with ulcers taking a large amount of calcium together with alkali, serum calcium rose as an exception. The intestine probably temporarily lost its control over calcium absorption while the kidney may also have failed to adequately excrete the extra calcium load. Under no other circumstances, not even a high oral calcium intake, have we seen such a rise in serum calcium.

What would happen if we took too much calcium over a long period of time? Intestinal calcium absorption would decrease to adjust to the high calcium intake, but some calcium would still "sneak in."

Bone is such a huge calcium reservoir that it could take up and store the excess calcium. Bone accounts for about 15 percent of the body weight, and a large part of the bone consists of calcium. At least 1 kg or 1,000,000 mg calcium is found in an adult body. If we compare this huge calcium pool in the bone with the daily calcium intake as food, we can immediately see the great difference. Even if we are taking 1,000 mg of calcium daily, only 30 percent or 300 mg is absorbed through the intestine into the blood stream. While it is diluted in 5 liters blood and circulates through the body, almost two-thirds or 200 mg is lost in the urine through kidney filtration. Only the remaining 100 mg or 1/10,000 the amount in bone is available to be deposited in the bone. Over a large surface of contact between the bone and blood, a much larger amount of calcium is constantly being exchanged. Although the actual amount of calcium exchanged between bone and blood has not been calculated directly, it may be as great as 100 times the amount absorbed from the intestine!

Even if you were to double your dietary calcium intake, the amount of calcium actually absorbed from the intestine would not increase proportionally because of the adjustment by the intestine. The increase from 100 mg to maybe 150 mg of absorbed calcium is very little compared to the vast calcium pool in the bone. The bone can easily take up this amount, and practically nothing is deposited in the soft tissues.

When even 0.1 percent of calcium comes out of bone, this is already 1,000 mg, 20 times as much as the increase of absorbed calcium. Calcium from bone is not different from calcium absorbed from the intestine, but the amount is much longer.

When calcium intake is sufficient, excess parathyroid hormone secretion is not necessary and calcium safely stays deposited in the bone.

"Do you not see that whatever goes into a man from outside cannot defile him, since it enters, not his heart but his stomach, and so passes on?" (Mark 7: 18–19)

When calcium intake is deficient, parathyroid hormone is secreted in larger amounts, triggering the release of calcium from bone. All over the wide interface between the bone and blood, bone resorption increases and calcium is withdrawn from the bone in large amounts. This calcium can flood the soft tissue, depositing itself in blood vessels and brain, contributing to the development of arteriosclerosis, hypertension and dementia.

6. How Do We Know We Are Taking Enough Calcium?

Even if calcium intake is deficient, serum calcium does not change, because our body sacrifices everything to keep the serum calcium level constant. Bone has an almost limitless supply of calcium, and parathyroid hormone brings out as much calcium as necessary to keep up the serum calcium level. Measurement of serum calcium tells us whether parathyroid hormone and active vitamin D are working adequately, but not whether calcium is deficient. Measurement of parathyroid hormone is a good but indirect method for evaluating calcium deficiency. Whenever calcium intake is deficient, serum parathyroid hormone is elevated. This is why serum parathyroid hormone rises with advancing age. Bone mineral measurement is a more direct method, because a long standing calcium deficiency always results in a decrease of bone mineral, even if the bone calcium store is quite large. A loss of 50 mg of bone calcium daily would result in a 15 gram calcium depletion each year, until 450 grams or almost one-half of the total body store of calcium is lost in 30 years. Even if the currently available methods of bone measurement are yet insensitive, they are certainly capable of detecting this amount of calcium loss from the bone (S. M. Garn, 1967). In both men and women, bone mass decreases with age, and this decrease is accelerated after middle age, especially in women.

Since X-ray pictures are not enough to detect subtle changes of bone mineral, various non-invasive or painless methods of bone mineral measurements have been devised. Single (SPA) and dual photon absorptiometry (DPA) and quantitative computerized tomography (QCT) are the examples. By using these methods, we are now capable of measuring bone mineral content in the forearm and spine to detect changes accompanying aging and osteoporosis. Effects of drugs used to treat osteoporosis are also adequately assessed by these methods. As with other diseases, it is important to detect osteoporosis as early as possible for effective treatment.

Fig. 37 Active absorbable calcium is very well absorbed even in vitamin D-deficient rats. Each diet contains 4% calcium.

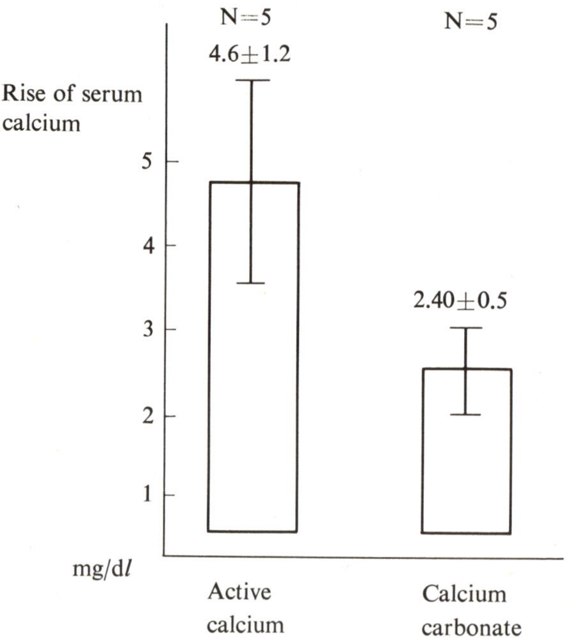

There are two kinds of bone: *compact* and *spongy*. Strong, hard compact bone is found in the *mid shaft cortex*—the outside cover of long, strong bones such as leg bones which support the body. *Trabecular* or spongy bone is found near the ends of the long bones and in the spine. It is softer with a more abundant blood supply. When the first single photon bone densitometer appeared, it was easier to meas-

Fig. 38 Active absorbable calcium made of oyster shell is more readily absorbed than calcium carbonate in a patient with postoperative hypoparathyroidism.

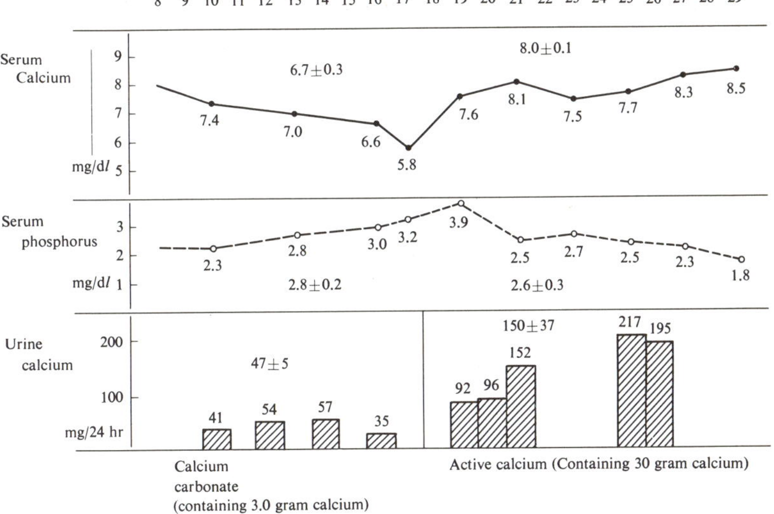

Fig. 39 Crystalline structure of active absorbable calcium (ionizable oyster shell) on the left, calcium oxide on the right.

ure the cortical bone in the middle of the forearm, because the bone there is hard, constant and not so variable. The error of measurement was small and the results were easily reproducible. This is still the standard method of bone densitometry. The cortical bone, however, does not rapidly change with age, in bone disease, or in response to treatment. When similar measurements are taken near the end of the forearm, it is possible to measure more spongy bone than cortical bone. Measurements of such trabecular bone have a wider variation, but they are more sensitive to bone mass changes with aging or disease states. Dual photon absorptiometry is like a double beam spectrophotometry. By using two different kinds of gamma rays emitted from gadolinium instead of only one kind of gamma ray emitted from iodine 125, interference by fat and other soft tissues is avoided. The dual photon bone densitometer has made it possible to measure the bone mineral content of the spine and hip, two bones deep within the body and most frequently affected by osteoporosis.

Computed tomography with phantoms for standardization also can accurately measure the density of the spongy bone in the spine. By computerizing absorption of X-rays at each part of the body, computed tomography makes it possible to reconstruct any part of the human body at any cross section. It thus becomes possible to take out a 1 cm thick section from the third lumbar spine, for example, to measure exactly how much X-ray is absorbed by all the tissue in this area. Therefore, mineral content of only the trabecular bone within the vertebral body may be measured, avoiding interference by cortical bone, spinal processes, calcified ligaments and blood vessels. It is usually difficult to avoid such interference in DPA. Correction for the fat content within the spongy bone is the only problem QCT faces and this may be solved by using two levels of X-ray energy.

These new techniques of bone measurement demonstrate increasing bone loss with advancing age. Cortical bone remains fairly constant until the middle age, while in females after menopause, bone loss increases and continues every year. Males lose bone at a slower rate than females. Spongy bone, on the other hand, is lost at a much earlier age, from about age twenty, in both males and females. These differences in the two types of bone underscore the importance of following the changes of both cortical and spongy bones this way. Bone measurement is the only way to detect calcium deficiency effectively in time to take preventive steps against serious bone loss. Neutron activation analysis is the most sophisticated way to measure the calcium content of the entire body. With this method, all calcium in the human body is temporarily transformed to a measurable form. This method is complex and has not yet been widely used.

It should be emphasized that none of these advanced methods of bone measurements should replace a good overall clinical evaluation. First of all, the normal range of each method of measurement should be established for each age and sex. Even after an individual is charted against the normal range of values, the physician should take a careful history of one's growth rate, childhood illnesses, dietary habits, occupation and all past diseases and injuries. A physical examination is also advisable to check for calcium deficiency. Laboratory tests are also important. Serum calcium should always be measured. Although this level is kept relatively constant, regardless of dietary calcium intake, any marked increase or decrease would indicate an abnormality of calcium metabolism. Serum phosphorus is also important because calcium and phosphorus have a high affinity for one another and frequently coexist in the bone and other places. The physician should also measure the amounts of calcium and phosphorus excreted in the urine, along with creatinine, a reliable and practically constant constituent of urine.

Other useful biochemical tests would include measurement of *alkaline phosphatase*, an enzyme produced mainly by osteoblasts; *osteocalcin*, a protein constituent of the bone, made by osteoblasts; and *hydroxyproline*, an amino acid which is found only in collagen and indicative of the degree of bone activity and degradation. Furthermore, by the technique of radioimmunoassay and receptor binding assay, it is now possible to directly measure the calcium regulating hormones, parathyroid hormone, calcitonin and $1,25(OH)_2$ vitamin D_3. Bone mass measurements together with these biochemical measurements will greatly facilitate an estimation of the calcium dynamics in the human body with a reasonable level of accuracy.

Summary

Calcium is not only an important component of our food, but also an element in our body which is indispensable to the maintenance of life. In order to keep ourselves healthy and enjoy longevity, adequate intakes of calcium are necessary to ensure optimum calcium balance, distribution between hard and soft tissue, and between inside and outside of the cell. The vast concentration differences between these compartments hold the secret of life. Life depends on the tension created by such profound differences in calcium concentration, which is the basis for the action of calcium as a messenger. Calcium is the fire of life. Let us keep this fire burning by achieving an adequate calcium intake to keep ourselves healthy.

References

Abe, E., Miyaura, C., Sakagami, H., Takeda, M., Konno, K., Yamazaki, T., Yoshiki, S., Suda, T.: Differentiation of mouse myoloid leukemia cells induced by 1 25-dihydroxy vitamin D_3. *Proc. Nat. Acad. Sci.* 78: 4990, 1981

Anderson, J. B., Barnett, M. S. R. E., Nordin, B. E. C.: The relation between osteoporosis and aortic calcification. *Brit. J. Radiol.* 37: 910–912, 1964

Arieff, A. I., Massry, S. G.: Calcium metabolism of the brain in acute renal failure. *J. Clin. Invest.* 53: 387–392, 1974

Avioli, L. V. (Ed): *The Osteoporotic Syndrome*, Grune & Stratton 1983

Borle, A. B.: Parathyroid hormone and cell calcium. *Excerpta Med. Int. Congr. Ser.* No 243 pp 484–491, 1972

Campbell, A. K.: *Intracellular Calcium*, John Wiley & Sons, 1983

Chen, K. M., Yase, Y. (Ed): *Amyotrophic Lateral Sclerosis in Asia and Oceania.* Shyan-Fu Chou, National Taiwan University Publisher 1984

Crawford, M. D.: Hardness of drinking water and cardiovascular disease. *Proc. Nut. Sci.* 31: 347–353, 1972

Deftos, L. J.: "Regulation of calcitonin secretion. Effect of species, age and sex," In: *Hormonal Control of Calcium Metabolism* (Cohn DV et al Ed) *Excerpta Med. Int. Congr. Ser.* No 511 pp 266–270, 1981

Elkeles, A.: A comparative radiological study of calcified atheroma in males and females over 50 years of age. *Lancet* 2: 714–715, 1957

Fujita, T., Okamoto, Y., Tomita, T., Sakagami, Y., Ota, K., Ohata, M.: Calcium metabolism in aging inhabitants of mountain versus seacoast communities in the Kii Peninsula. *J. Amer. Geriat. Soc.* 25: 254–257, 1977

Fujita, T., Sakagami, Y., Tomita, T., Okamoto, Y., Oku, H.: Insulin secretion after oral calcium load. *Endocrinol. Japon.* 25: 645–648, 1978

Fujita, T.: Serum calcium—an important determinant of the survival of elderly patients. *Kobe J. Med. Sci.* 30: 1–6, 1984

Fujita, T., Okamoto, Y., Sakagami, Y., Ota, K., Ohata, M.: Bone changes and aortic calcification in aging inhabitants of mountain versus seacoast communities in the Kii Peninsula. *J. Amer. Geriat. Soc.* 32: 124–128, 1984

Fujita, T.: Aging and Calcium, *Mineral and electrolyte metabolism* 12: 149–156, 1986

Garn, S. M., Rohmann, C. G., Wagner, B.: Bone loss as a general phenomenon in man. *Fed. Proc.* 26: 1729–1736, 1967

Goldstein, D. A., Chui, L. A., Massry, S. G.: Effect of parathyroid hormone and uremia on peripheral nerve calcium and motor nerve conduction velocity. *J. Clin. Invest.* 57: 88–93, 1978

Hossain, M., Smith, D. A., Nordin, B. E. C.: Parathyroid activity and postmenopausal osteoporosis. *Lancet* 1: 809–811, 1970

Imura, H., Nakagawa, S., Goto, Y., Kosaka, K., Sakamoto, N., Kaneko, T., Mimura, G.: "Complications of diabetes mellitus with special reference to diabetic osteopenia," In: *Internal Medicine* (Elsevier) pp 363–368, 1986

Kalu, D. N., Hardin, R. H., Cockerham, R., Yu, B. P.: Aging and dietary modulation of rat skeleton and parathyroid hormone. *Endocrinol.* 115: 1239–1247, 1984

Kanematsu, S. *Food & Nutrition* (in Japanese) 6: 135–147, 1953

Kaplan, L., Katz, A., Ben-Issac, C., Massry, S. G.: Malignant neoplasms and parathyroid adenoma. *Cancer* 28: 401–407, 1971

Knox, E. G.: Ischaemic heart disease mortality and dietary intake of calcium. *Lancet* 1: 1465–1467, 1973

Lansing, A. I., Alex, M., Rosenthal, T. B.: Calcium and elastin in human arteriosclerosis. *J. Gerontol.* 5: 112–119, 1950

Okano, K., Fujita, T., Orimo, H., Yoshikawa, M.: Age and calcium metabolism in relation to cardiovascular changes induced by renal injury with sodium sulfacetylthiazole. *J. Amer. Geriat. Soc.* 18: 458–470, 1970

McCarron, D. A., Lucas, P. A., Sckneidman, R. J., LaCour, B., Drüecke, T.: Blood pressure development of the spontaneously hypertensive rat after concurrent manipulation of dietary Ca^{2+} and Na^{+}. *J. Clin. Invest.* 76: 1147–1154, 1985

McCarron D. A., Morris, C. D., Cole, C.: Dietary calcium in human hypertension. Science 217: 267–269, 1987

Rixon, R. H.: Parathyroid hormone. A possible initiation of liver regeneration. *Proc. Soc. Exp. Biol. Med.* 141: 93–97, 1972

Rixon, R. H., Whitfield, J. F.: The radioprotective action of parathyroid extract. *Int. J. Rad. Research* 3: 361–367, 1961

Yocowitz, H., Fleischman, A. I., Bierenbaum, M. L.: Effects of oral calcium upon serum lipids in man. *Brit. Med. J.* 1: 1352–1354, 1965

Yase, Y., Yoshimasu, F., Uebayashi, Y., Iwata, S., Kimura, K.: Amyotrophic lateral sclerosis. Interaction of divalent metals in CNS tissue and soft tissue calcification. *Proc. Japan Acad.* 50: 401–406, 1974

Index